Make Your Own Electricity

Make Your Own Electricity

Terence McLaughlin

David & Charles
Newton Abbot London North Pomfret (VT)

First published 1977
Second impression 1977
Third impression 1978

ISBN 0 7153 7109 6

Set in 11 on 13 point Baskerville
and printed in Great Britain
by Biddles of Guildford
for David & Charles (Publishers) Limited
Brunel House Newton Abbot Devon

Published in the United States of America
by David & Charles Inc
North Pomfret Vermont 05053 USA

Contents

1 What kind of supply? 7
2 Electrical units 10
3 Power requirements 21
4 Generators 24
5 Engine-driven generators 45
6 Wind generators 50
7 Power from water 71
8 Power from the sun 75
9 Thermoelectric power 81
10 Storage batteries 86
11 Inverters and converters 100
12 Changeover switches 108
13 Safety 111
Appendices 117
Index 125

1 What Kind of Supply?

Electricity is essential for modern life, and the measure of our dependence on it is the chaos caused by any interruption in the supply, the outcry when it rises in price. Fifty years ago a power cut meant merely that people got out the candles or oil lamps and waited until the lights came on again, but now any prolonged failure means passengers stuck in lifts, traffic in confusion without control lights, factory machinery dead, and offices unable to function because even the typewriters need electricity. In the home the darkness is the smallest problem: the central heating goes cold because the circulation pumps and controls are out of action; the deep-freeze and refrigerator start to thaw, with a possible loss of valuable food; and contact with the outside world breaks down with the failure of radio and TV.

Accepting that we cannot do without it, what we can do to buffer ourselves against the increasing cost of electricity and the inevitable breakdowns in supply caused by accidents, overloading, strikes and other causes?

It is quite feasible to have at least some control over your electricity supply. First decide exactly what you want in the way of an alternative supply. There are four main possibilities:

Emergency supply You plan to use the ordinary mains (grid) supply most of the time, but you find some way of running the essential equipment—some lights, radio, TV, central-heating pumps, any heaters for fish tanks, the deep-freeze and refrigerator, an electric razor, perhaps an electric blanket or two—during power cuts.

Partial replacement You do not plan to cut yourself off from the public supply, but you organise some alternative source of power to run some of your equipment at lower cost. This benefits you financially and helps to solve the general energy-supply crisis.

Separate system Some of your equipment is wired separately from your ordinary house supply—perhaps you have a greenhouse that needs electricity to operate moving lights, a water pump, or soil-heating cables, or a workshop or shed separate from the house, an incubator for chickens, and so on—and you realise that this would be the ideal place to instal a limited alternative supply entirely separate from your house needs.

Complete replacement You want to give up the use of public electricity altogether, or perhaps you live in an area where the cost of getting it to you is prohibitive: you therefore think in terms of a complete supply system to cover all your electrical needs.

The first three possibilities are fairly easy to achieve, without a crippling capital outlay and with no more than the normal handyman's type of skill, and this book will give a number of systems for achieving success with an alternative supply. The fourth possibility, complete independence from the mains supply, will cost you a substantial sum even if you are prepared to do all the work yourself, because designing, installing and maintaining a full-scale electrical plant suitable for the needs of the modern house is a big job.

The second decision to make is whether to run your plant from conventional fuels such as petrol or diesel oil in an engine, or whether to make use of the vast natural resources of the sun and wind. Conventional fuels give you a steadier output of electricity and, if you use the commercially made plants, will supply all your electrical needs with little maintenance work. On the other hand, you have to buy the fuel and you will not be helping the energy shortage. Wind power is the most attractive of the alternative methods, and in Chapter 6 you will find details for a number of generator systems that make best use of this free resource. However, wind-driven generators do need a lot more attention than the normal petrol-driven generating set, and you will probably not be able to build one large enough to supply all your electrical needs unless you have a fair amount of engineering knowledge and a well-equipped workshop. A very large wind-driven generator is rather an awesome thing, and needs to be treated, and designed, with proper respect for the safety of the structure. If you build one of the medium-sized wind generators described in this book, however, you can always add another one to supplement your supply. The experience with one may suggest improvements to try on another

Some quite detailed sets of instructions are included here for readers who prefer to work closely to a plan. Modifications are suggested for the reader who prefers to take a general idea and work out the details in his or her own way; and some of the newer methods of generating electricity which are not yet practicable for domestic supply, but may become important in the near future, are noted.

For all our talk about energy crises, there is no real shortage of energy at all: it is just that we choose to employ a very limited range of generating methods, mainly from fossil fuels (coal and oil). The sun radiates millions of times more energy than we could ever use, and even the fraction that penetrates the clouds and our polluted atmosphere is hundreds of times more than our most extravagant requirements. Unused wind energy is vastly greater than the output of all the power-stations in existence. Nuclear energy now seems far less attractive than it did in the 1950s, and the scientific arguments over light-water reactors and so on remind us that nuclear power-stations can be unpleasant neighbours.

This book may therefore at least encourage some people to make their own contribution to a better and more rational solution to the 'energy crisis', and at the same time make themselves a little less vulnerable to the economic and political pressures that can bedevil the energy industry.

2 Electrical Units

Most of the projects and suggestions in this book do not require any profound knowledge of electrical or physical theory (indeed, very few professional electricians have this sort of knowledge), but a few elementary matters should be understood if only so that you can adapt the ideas here to your particular needs and circumstances. The following definitions and formulae may help to make later portions of the book clearer for those readers unfamiliar with electrical work.

Volts Voltage is basically a measure of the pressure with which electricity is supplied. The higher the voltage, the more easily the electricity will be able to pass through materials. For example, a torch (flashlight) battery gives about 1½ volts (V) and, as you may have noticed, this can easily be cut off from the bulb by something as simple as a patch of rust on the switch or some grease on the contacts. On the other hand, the public supply gives 240V in the UK (120V in the USA), and this is not interrupted by anything less than a substantial thickness of plastic, paper, etc.

Low voltages are safer to handle because they cannot penetrate the skin: below about 50V there is very little danger of a dangerous shock, but a voltage much above this can kill (see pp 111–13). As voltages go up, it becomes more difficult to stop the electricity flowing, even when we want to—even the mains voltage can jump across a gap momentarily, causing a spark (you often see this if you switch off something in the dark, or pull out a plug from a power point), and high voltages such as are used in the grid distributing lines can jump nearly 10in (25cm), which is why the pylons have to be so high and the cables so widely separated.

The following table will give some idea of the properties of electricity at various voltages.

Table 1 Effects of Various Voltages

Source	*Voltage*	*Properties*
Single torch battery	$1\frac{1}{2}$	Easily interrupted by grease, corrosion, etc
Transistor-radio battery	9	Similarly needs good contacts
Car battery	12	Contacts can lose power if not scraped clean
Heavy truck battery	24	Not so sensitive to dirt, etc
Low-voltage outdoor supply	50	Can cause a very mild shock, does not leak when contacts are wet
US public voltage	120	Can cause a shock but not as serious as UK mains
UK public voltage	240	Causes serious shocks, needs careful insulation, can spark across small gaps
UK factory (3-phase)	440	Causes fatal shocks, needs even more careful insulation, can arc (long-continued sparks)
UK substations	33,000	Sparks can jump 1cm in air
UK grid (old sections)	132,000	Sparks can jump 5cm in air
UK grid (new sections)	400,000	Sparks can jump 25cm in air

It goes without saying that any contact with, or even close proximity to, the last three voltages causes instant death. That is one of the reasons why the 400,000V pylons are 164ft (50m) high.

To save writing noughts when dealing with such voltages, they are usually expressed in kilovolts (kV = 1,000V). At the other end of the scale, some of the devices described in this book produce very low voltages, and have to be strung together to make a satisfactory supply. Very small voltages are measured in millivolts ($\text{mV} = \frac{1}{1{,}000}\ \text{V}$).

Amperes show the amount of electricity passing along a wire or through a piece of equipment. This can vary from the few thousandths of an ampere (A) which pass from a battery into a transistor radio or similar electronic device (milliamperes,

$mA = \frac{1}{1,000}A$) to the 450A or so that pass from a car battery to the starter motor when you first switch on, or the thousands of 'amps' in some industrial welding processes. The most important consideration with amperes is that the more a wire has to carry, the thicker it must be, just as you need a wide pipe to carry a lot of water. Ordinary flex as used for hanging light fittings and similar purposes will carry a maximum of about 5A safely; the heavy cable used for ring-mains can carry 20A. The lead to a car starter-motor, which has to take a load of as much as 450A for a short time, is a really heavy mass of copper, and similar very heavy cables are used for welding and other jobs involving large amperages.

If you try to force too many amps through a wire, by connecting it up to some piece of equipment that takes a far greater current that the wire is rated to carry, the wire will get hot. This is the cause of a great many fires in industry and in the home: very often a power point, or even a light fitting, is festooned with multiple adaptors and two-way switches until it is supplying eight or ten electrical gadgets. Most light fittings are not designed to supply more than 5A, although they can usually supply 10A for a short time without trouble. If, however, they are regularly supplying 15A then something will get hot enough to burn. It may be the actual fitting, but it can equally well be a piece of the connecting cable under the floorboards or behind the skirting board, where it will not be noticed until it sets fire to the wood.

You can see the effects of putting too high a current through a thin piece of wire by dropping a piece of steel wool over the contacts of an ordinary torch or lantern battery. The steel wool acts as a bunch of very thin wires, able to carry a fraction of an amp but not much more, and the heavy current from the battery will rapidly make it red-hot: it will eventually burn away.

Fuses are devices which make use of this effect—they are very easily melted by heat, and if the circuit in which they are placed is overloaded, so that too many amps are running through it for safety, the fuse is the first piece of wire to burn. It melts and cuts off the electricity from the circuit. The advantage is that you know the point at which overheating will occur, if it is going to (ie, in the fuse-box). If you replace fuses with hairpins, pieces of wire, or even fuse-wire which is too heavy for the rating of the circuit, you simply go back to the system where overheating can occur anywhere in the circuit, usually where you cannot see it until a fire breaks out.

Watts are a measure of the actual power used by a piece of apparatus—this can vary from one watt (W) or less used by some pieces of electronic equipment (calculators and so on) to thousands of watts used by large heaters, cookers, etc.

Because the range of apparatus is so great we also use the units *milliwatts* (mW $= \frac{1}{1000}$ W), *kilowatts* (kW = 1000W), and, for the output of power stations, *megawatts* (MW = 1,000,000 W). Chapter 3 tells you the average wattage of most pieces of household equipment. (Just to get these figures in scale,. the peak output of the UK generating system is about 56,000MW (56,000,000,000W), or about 2.3kW per household—the rest is used in industry.)

Wattages are easy to calculate, as the watts used by a piece of equipment are equal to the voltage multiplied by the amperage passing through: W = VA.

Very often, however, we know the wattage, because manufacturers print it on the equipment, and we know the voltage if it is the standard mains voltage, so we can use the same formula to find out the amperage passing along the wires to the equipment.

For example, consider a 100W electric lamp. The mains voltage in the UK is 240V, so $100 = 240 \times A$, or A = 0.42 amps. This shows that the 5A cable usual for lighting fixtures can easily carry about ten 100W lamps without overloading. Now consider a 3-bar electric fire—most of these have 1000W bars, so in this case $3{,}000 = 240 \times A$ or A = 12.5 amps. Most modern power points are rated to carry 13A, so a 3-bar fire is about as much as can safely be run from a point.

This rule has important consequences for home-produced electricity. Many of the systems described in the book give a supply of lower voltage than 240, sometimes as low as 12V. Just as many watts are needed to run a lamp, motor, or any other piece of equipment at 12V as at 240V, so to make up for the low voltage the amperage must be much higher. For example, if you decide to make up emergency lighting with 12V car headlamps, and you want a reasonable amount of light, you will probably use several 45–50W lamps. If you had, say, ten 50W lamps run from the 240V mains, the total amperage would be just over 2A, but at 12V the amperage must be nearly 42A, which means using very thick cable.

The same principle applies if you are taking low-voltage power from a source such as a wind generator (see Chapter 6). This need for heavy cables can add expense to setting up an alternative electricity system using low-voltage sources, and Chapter 10 gives details of devices to convert low-voltage supplies to 240V or 120V.

For many mechanical purposes, such as driving motors or pumps, or using solenoids to open a door or operate greenhouse ventilators, it is convenient to have a relationship between electric power in watts and mechanical power in *horsepower:* very often motors are rated in HP instead of W. The relationship is fairly simple: 1HP = 746W.

However, owing to inefficiencies in the conversion of electric power, by motors in particular, it is necessary to supply considerably more than 746W for every 1HP output you need: a typical '¼-horse' motor, for example, actually takes from the mains 1.8A at 240V, or 432W, giving an equivalent of 1728W per HP. Larger motors are designed to be rather less inefficient, but you still need about 12–1400W per HP. These losses are partly due to friction and similar losses in the motor and partly to a systematic loss in efficiency called *power factor* that affects most equipment run from the ac mains, especially if it has coils as used in motors and solenoids. Calculation of power factor is beyond the scope of this book: you simply have to remember that you need to put into a motor about twice the power you get out of it.

Watt-hours represent the total energy used by a system. Obviously if you leave an electric fire on for ten hours it uses more energy than if it were on for one hour, and the measurement of this energy is the *watt-hour,* or (for most equipment) the kilowatt-hour (kWh), which is the amount of energy used by a 1,000W appliance in one hour, or a 2,000W appliance in half an hour, and so on. The kWh is also called simply a *unit;* these units are of course the figures shown on your electricity meter and adding up to such alarming amounts on your electricity bills.

Heating engineers in the UK and USA often use another measurement of energy, the *British Thermal Unit* (BTU), which is based on the British system of weights and measures and the Fahrenheit thermometer, while kWh are metric. The relationship between the two units is that 1BTU = 0.000293kWh. If your central-heating expert, or the table of recommended values you are using, says that your living-room needs an average output of 3,000BTU per hour (72,000BTU per day), this is equivalent in electrical terms to 21.1 units per day, while you could provide from, say, a 2kW radiator run for just over ten hours per day, or a 1kW radiator left on most of the time. Similarly, if you have a solid-fuel boiler and think of replacing it by electric heating, look at the rating of the boiler in BTU. If it is producing an average of 30,000BTU per hour over the whole day, you will have to think in terms of paying for about 21 units per day for heating.

Ohms measure the resistance of a piece of equipment, length of wire, or almost anything that electricity has to pass through. Some materials like rubber and most plastics present so much resistance to the passage of electricity that they can be used to hold it in—they are used for covering wires and switches and so on. They are *insulators*, and their resistance is measured in thousands or millions or ohms (Ω). The units are called *kilohms* (k = 1,000Ω) and *megohms* (M = 1,000,000Ω). High resistances like this are also often found in electronics work.

At the other extreme, metals like copper and aluminium allow electricity to pass so readily that they are used as *conductors* to transmit power from place to place. Their resistances in bulk are measured in millionths of an ohm, *microhms* ($\mu\Omega = \frac{1}{1,000,000}\Omega$), and even when they are stretched out into thin wires the resistance is usually only a few ohms per 1,000 metres of wire.

If you know the voltage in use and the current or wattage of a piece of equipment, you can easily find out its resistance. *Ohm's Law* states that Resistance in ohms (R) = Voltage in volts (V)/Current in amps (A) so, for example, a 60W bulb that takes a current of $\frac{1}{4}$A at 240V has a resistance of $\frac{240}{\frac{1}{4}} = 960\Omega$.

If you know the wattage you can take advantage of the fact that W = VA, or $A = \frac{W}{V}$; this gives $R = \frac{V^2}{W}$ so that the resistance of a 3kW electric radiator at 240V is $\frac{240^2}{3000} = 19.2\Omega$.

You need to know something about the resistances of electrical materials for two main reasons if you are thinking of setting up your own supply. The first and most important reason is that wires and cables with a very low resistance per yard of length can build up quite a significant resistance over long runs of cable. If you have a low-voltage supply, it can be seriously depleted by the wrong choice of cable.

For example, the standard cable used for domestic ring-mains wiring is so-called $\frac{1}{1.78}$ where each conductor has a cross-section of 2.5 mm^2. (Fortunately the cross-sectional area is being used more and more to identify cable, which makes it a lot easier to find out if a particular grade is suitable for your purposes). This cable has a resistance of 0.008Ω per metre, taking each conductor, or 7.3Ω per 1,000 yards; as in almost every case there will be a return wire of exactly the same length we can call the resistance 'there and back' 0.016Ω per metre of connecting cable.

As such cable when used in the home, with a 240V supply, can carry enough current to supply about 5kW load, you might think that it would be adequate to carry a 12V supply from a wind generator, for example, for a distance of 25 metres to feed a system of 6 × 40W car headlamps. Unfortunately you will find in practice that the resistance of the cable, $25 \times 0.016 = 0.4\Omega$, is almost as great as the resistance of the lamps, about 0.6Ω, so instead of getting 12V at the house you will find only just over 7V, and instead of the 240W they need your lamps will get just over 86W, hardly enough to make them glow properly.

Obviously much thicker cable is needed to carry low voltages. You can calculate the effects of various cables by the following method:

1. Measure the length of cable from your source to the load, in metres if possible, or measure in yards and multiply by 0.9144.
2. Consult Appendix 1, which gives resistances per metre for various sizes of cable. Choose a likely cable and multiply your distance in metres by the $\frac{\Omega}{m}$ to get a total resistance. Call this R.
3. Measure or calculate the resistance of the load (lamps, motor, or whatever else). If you know their rated wattage W at V volts, then the resistance in ohms $= \frac{V^2}{W}$ Call this resistance R'.
4. The actual voltage across the load will be $\frac{VR'}{R'+R}$ and the true wattage attained will be $\frac{V^2R'}{(R'+R)^2}$

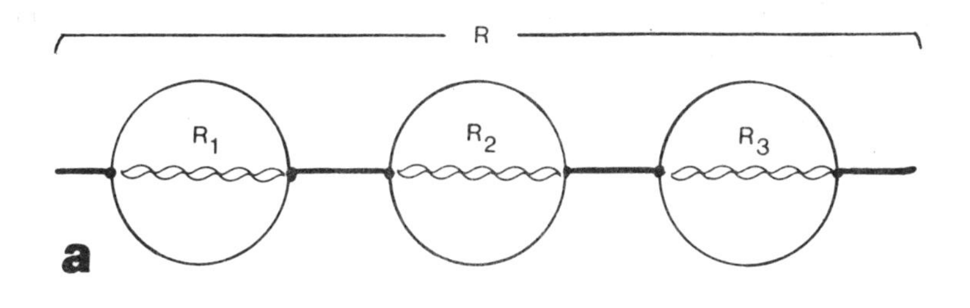

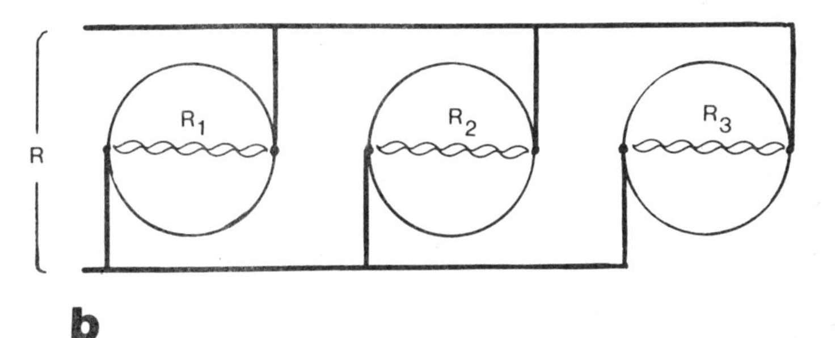

1 Lamps in series (a) and parallel (b). When in series the total resistance $R = R_1 + R_2 + R_3$: when in parallel
$\frac{1}{R} = \frac{1}{R_1} + \frac{1}{R_2} + \frac{1}{R_3}$

If the actual voltage is going to be more than about ten per cent lower than the voltage at the source, choose a larger size of cable. For example, with the figures we have just considered, 25m of cable feeding a set of lamps with resistance 0.6Ω, $10mm^2$ cable would give a voltage at the lamps of just over 11V, a satisfactory level.

If you have several lamps or similar pieces of equipment that are connected together to make use of the power, Fig 1 shows how to assess the total resistance. If they are connected in *series* (*Fig a*) the total resistance is given by $R = R_1 + R_2 + R_3 + R_4 + R_5 \ldots$ and so on. If they are connected in *parallel* (*Fig b*) the total resistance is given by

$$\frac{1}{R} = \frac{1}{R_1} + \frac{1}{R_2} + \frac{1}{R_3} + \frac{1}{R_4} + \frac{1}{R_5} \ldots \text{ and so on.}$$

The other reason for needing to know about resistances is to protect your prime power source, or the batteries you run from it. If you connect something with a very low resistance to your mains supply, a large current flows, but within a fraction of a second the fuse blows and cuts off the supply. If you put a piece of wire or some similar unsuitable thing in your fuse-box, you may blow a company fuse (which will cost you money to have replaced), but even then you will never be in any danger of damaging the company's generators, because their capacity is so large compared with any possible load you could put across the mains.

With your own supply, however, you have a strictly limited capacity for overload, and if you connect an appliance with too low a resistance to your system a very heavy current will flow not only through the appliance and wiring, but through your generator and/or batteries as well, probably burning out your generator and ruining your batteries. If, for example, you short-circuit an ordinary car battery by putting a screwdriver or other solid metal bar between the terminals, currents up to 400A will travel through the screwdriver and return through the cells of the battery, generating heat at the rate of 4.8kW. Ten minutes of this will be enough to bring the battery to boiling point, and meanwhile the lead plates will be buckling in the heat until they touch and short-circuit internally, rendering the battery permanently useless. Keep a careful eye on connections, therefore, and always include a properly rated fuse in your circuits.

Direct current and *alternating current* are two ways of distributing electricity. With direct current (dc) the electricity always flows in the same direction round a circuit, usually at a steady voltage and amperage. Batteries supply dc, and it is used in nearly all electronic equipment (although the actual external supply may not always be dc).

Alternating current (ac) travels one way round the circuit, then stops and comes back the other way. The number of these *cycles* per second is known as the *frequency,* and is expressed in Hertz (Hz), or occasionally as cycles/second (c/s): the usual frequency in the UK is 50Hz, that in the USA 60Hz. Quite often objects like lamp filaments vibrate in sympathy with the frequency, so that you can actually hear them—50Hz is quite near the musical note G_1, 60Hz around B_1. More annoying is the fact that the frequency sometimes gets through electronic equipment driven from ac mains, and appears at the loudspeakers of radios, record players, and so on as a 50- or 60Hz hum underlying all the other sounds. If you have a battery/mains radio you may have noticed, unless it is a very well designed one, that it hums when run from the mains but not when run from the batteries. This is because the battery is dc and the mains ac.

Because the voltage and current are constantly changing, building up from zero to a maximum, then falling off and eventually reversing, it is difficult to say exactly what the voltage or amperage of an ac supply is. The mains in the UK fluctuates from about 340V to —340V, going through 0V in between, but it stays at this *peak voltage* for so little time that you would get a large overestimate of the wattage if you used 340V in calculations. What is actually used is the *root mean square* (rms) voltage, which is 0.707 times the peak voltage (240V in this case), and rms current in amperes. Multiplying these two rms values gives the true wattage supplied by the alternating current, so that calculations can be carried out exactly as for dc systems.

So far it might seem that the use of ac involves nothing but trouble, and it is quite true that ac has very few advantages over dc for the consumer. Where its advantages lie is in the production and distribution.

Alternators, which produce ac, are simpler to construct and

need less servicing than dynamos, which produce dc. This will be considered in more detail in Chapter 4 when deciding which type of machine to use for home generation. The overwhelming advantage of ac, however, is that the voltage can be altered, with almost no loss of power, by *transformers*. We have seen several times that there are advantages in running systems at as high a voltage as possible, so as to reduce the current in amps and get more power through a cable of a given size. National distribution grids use voltages of as much as 400,000V (trials are going on with an 800,000V system), so that enormous amounts of power can be distributed through relatively small cables with very low losses. However, such a system would be useless if without an easy and efficient way of bringing the voltage back to a reasonable level for the consumer. Transformers do this with hardly any loss.

There are devices for stepping up the voltage of dc supplies, and bringing them down again for the consumer, but so far these are less efficient than the transformer used with ac.

The transformer can be used for many similar purposes in domestic supplies: to reduce the voltage from 240V to about 24V for hedge clippers and lawn mowers, for example, so that they are safer when used outdoors, or to step up the voltage from 240V to the thousands of volts needed for the picture tube of a TV set.

In Chapter 10 there are some suggestions for using transformers to convert a low-voltage primary supply to 240V so that it can be used to run ordinary electrical equipment designed for mains use: this requires the conversion of dc to ac so that the transformer can be employed.

3 Power Requirements

Whether your alternative power supply is to be used only for emergency services, or for regular boosting of your electricity supply, you will need to go round the house and list the wattages of all your equipment, especially the items that you could not readily do without during a power cut. List essential lights, deep-freeze, refrigerator, central-heating circulation pump, the fan on the boiler (if it has one), electric kettle, some kind of cooking facility (if only one ring or plate), electric blankets, any essential baby equipment, electric razor, radio and TV, heating to fish tanks and other equipment that cannot be left off for long without serious consequences, and any thing that you need to carry on something like normal life. If you run a business from home, for example, your electric typewriter or similar office machinery may be essential.

For a supplementary system, designed to take some of the load, or to supply a greenhouse, workshop, or some other outside building, you need list only the items to be supplied, but remember that it may be an advantage to be able to use your installation for emergencies such as power cuts. It would be very galling to install an efficient alternative system in a workshop, say, and then find during power cuts that your workshop was a blaze of light while your house was dark and cold.

Most pieces of electrical equipment have their wattage marked on them, but the following table may help in sorting out the power requirements where these are not immediately obvious. When totalling, remember that in fact you will rarely be using all your equipment at once. The values quoted, of course, may differ from the ones specified for some pieces of equipment.

Table 2 Wattages of some domestic equipment

Equipment	Watts	
Air conditioner	1,000–1,500	
Boiler fan	250–600	
Blanket	50–200	
Bottle sterilizer	400–550	
Central-heating pump	50–80	
Clock	1–10	
Deep-freeze	300–800	
Dishwasher	500–1,200	up to 2,000 when heating water
Table fan	25–75	
Floor polisher	220–400	
Food mixer	50–200	
Hair dryer	200–1,200	usually 50–150 when running cold, 500–1,200 when hot
Iron	500–1,000	
Kettle	2,000–3,000	
Lights (filament) as marked, usually	40–200	
Lights (fluorescent) as marked, usually	40–80	
Radio	40–100	
Razor	5–15	
Record player	100–250	
Refrigerator	200–300	
Sewing machine	30–100	
Television (portable)	20–50	
Television (large colour)	200–250	
Toaster	400–800	
Typewriter	100–200	
Vacuum cleaner	250–500	
Washing machine (twin-tub)	250–600	up to 3,000 when heating water
Washing machine (automatic)	250–500	up to 3,000 when heating water

Heating equipment usually takes the heaviest current: th normal bar type of electric fire takes 1,000W per bar, fa heaters are usually 500–1,000W, and night storage radiator 2–3,000W. Depending on the size, an electric cooker may tak up to 6,000W when all the plates and other elements are on.

For motor-driven equipment, a single assessment may b made from the size of the motor: in general, induction motor (as used in all small ac equipment) need a heavy current t start them against load, and then settle to a smaller consumption when they are running. The approximate figures are:

Table 3 Motor wattages

Motor HP	*Starting watts*	*Running watts*
1/6	900	200
$\frac{1}{4}$	1,300	300
$\frac{1}{3}$	1,500	360
$\frac{1}{2}$	2,200	520
$\frac{3}{4}$	3,400	775
1	4,000	1,000

Farms and smallholdings often suffer greatly when power cuts occur, and badly need an emergency power source. Table 4 shows the approximate demands of some agricultural activities.

Table 4 Power demands in agriculture

Application	*Approximate wattage*	*Power demand (kWh)*
Feed grinding	1–10,000	1 per 200 lb feed
Grain drying (heated air)	3–10,000	6 per ton grain
Grain elevating	300–3,000	4 per 1000 bushel
Hay curing (heated air)	3–7,000	50 per ton
Milking machine	500–2,000	$1\frac{1}{2}$ per cow per month
Silo unloading	300–500	$\frac{1}{2}$ per ton
Brooder	125–1,000	0.8 per chick per month
Incubator		0.2 per egg per month

4 Generators

In 1821 Michael Faraday discovered that if he moved a ring of wire in the field from a magnet, a small but detectable electric current passed through the wire. Over the next few years he proceeded in his unostentatious way to set down all the principles that anyone has needed since to understand electromagnetism. But he did not invent the generator.

This has often been quoted as one of the great historical opportunities lost, but in fact the society of the time could have

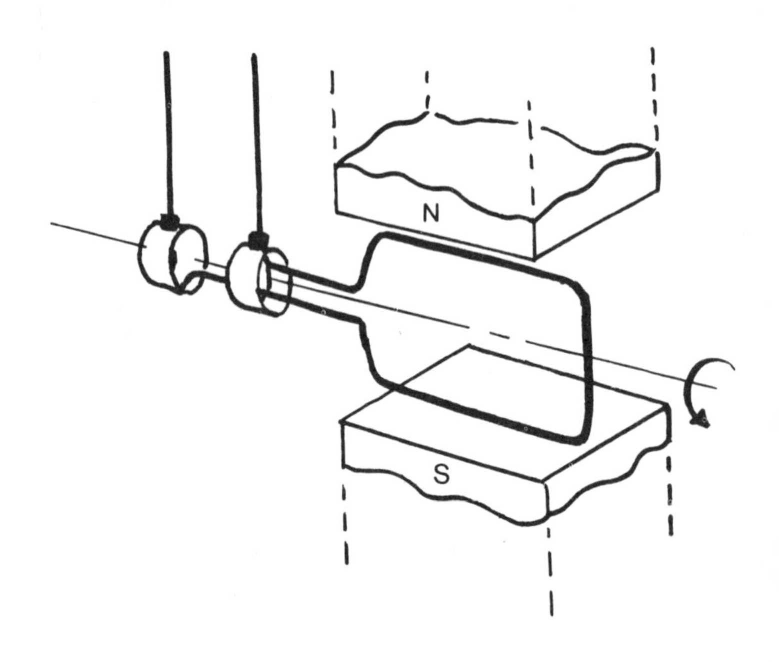

2 Elementary ac generator

done very little with the generator even if Faraday had invented it. There were no electric lamps, no electric motors, no radios or TVs—electricity was just something that scientists played with, and as Faraday was quite capable of making fundamental electrical discoveries with a battery made of seven halfpennies and some old pieces of zinc, he himself needed no elaborate generating systems.

Figure 2 shows the principle of the simplest type of generator, which will serve to explain all the important points about more complex ones. If a coil of wire is rotated between the poles of a magnet, it will produce an alternating current, at its maximum as the wire passes close to the magnet poles, at zero when the wire is horizontal, and reversing as the wire passes the opposite pole. If the wire is rotated fifty times a second, the frequency of the ac produced will be 50Hz; the voltage and current produced depend on the strength of the magnetic field and the number of turns of wire passing the magnets every second. Practical generators have dozens or even thousands of turns of wire in the coils, not just the one shown in Fig 2.

The fact that the power depends on the number of turns of wire passing the magnets gives two choices for running the generator: we can either have very large coils with many turns moving fairly slowly, or have smaller coils moving faster, to get the same power. To use large coils creates some difficulties because the rotating coils (the *armature*) become very heavy and costly to make, and the coils take up a lot of space between the magnet pole-pieces, so the magnets have to be large as well. On the other hand, a smaller armature turning more rapidly will produce current at a higher frequency, which may not be desired. For a 50Hz supply the maximum is 3,000rpm, which is in fact used in many turbine-driven generators.

In practice the slower-moving large armature is used extensively for large commercial generators, where weight does not matter very much, but the mechanical problems of having such a large mass of metal revolving fast are very great. To allow even slower speeds than fifty turns per second (3,000 rpm) it is usual to put several sets of magnet poles round the circle, so that the frequency is multiplied: with four poles, for example, rotation at twenty-five turns per second (1,500rpm) still gives 50Hz supply, and a common design for large generators providing commercial supplies involves a 28-pole *stator* (the fixed part) and an armature revolving at 214.3rpm (which gives 3,000 cycles per minute or 50Hz when passing the fourteen sets of pole-pieces).

Smaller generators, where portability is more important, are usually made to run faster so that the armature can be lighter and the total dimensions of the generator smaller. Such generators are used for car dynamos and alternators, which are often rated to run at speeds up to 9,000rpm or more.

DC generators

For many purposes a generator that supplies dc rather than ac is convenient. This can be achieved by a simple modification of

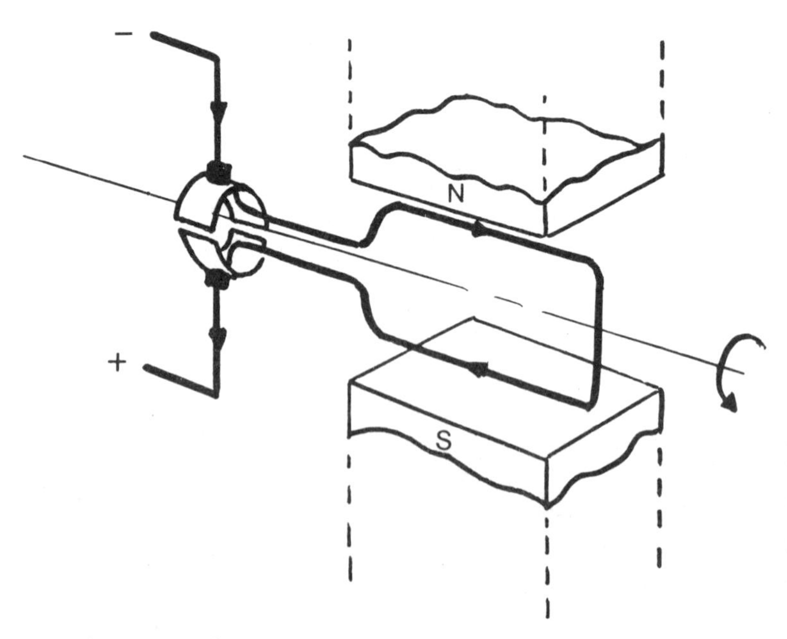

3 Elementary dc generator

the basic design (*see Fig 3*). The coil still revolves between the pole-pieces of a magnet, but the collecting ring (the *commutator*) is arranged in two halves, so that just as the direction of the current is going to reverse, it is transferred to the other terminal of the output leads. In this way, although the current goes from zero to its maximum twice in each cycle, the direction of the current coming out is always the same.

The variation in the value of the current can be smoothed out by having several poles, as in other large generators, and most practical dc generators have eight poles or more and a corresponding number of windings on the armature. This brings with it the problem that the commutator has to be divided into as many parts as there are poles, and with a multipole machine these segments can get very narrow. This can lead to trouble if there is any arcing or other electrical leakage between the various segments, which may be only a fraction of an inch apart. The segments are insulated from one another by thin slices of non-conducting material, usually mica, and the mechanical and design problems of turning this sandwich of copper conductors and mica slices into a smooth-running cylinder are considerable. The current is picked up from the conductors of the commutator by *brushes,* smooth pieces of graphite pressed against the revolving surface by springs, and these again need care and attention if they are to run smoothly and not spark or fail to contact some of the segments.

Revolving-field generators

As long as the magnet and the wire move relative to one another, it does not really matter whether the coil of wire is the rotor and the magnet the stator, or the other way round. An early design of car generator had permanent magnets revolving with the wheels just inside a fixed coil of wire: as the magnets moved an alternating current was generated in the coil and could be collected from the ends.

Such an arrangement has obvious advantages, as there is no commutator or slip rings, and the coil can be made as large as we like without any mechanical problems of making it move. The idea was not entirely successful because of the weakness of the permanent magnets available at the time, but the principle has many applications with newer types of generator which will be considered in the next section.

A *revolving-field* generator, in which the magnets move rather than the coil, can only produce ac, as there is no way in which the current can be reversed at each cycle as with the commutator revolving with the coils.

Improving the magnetic field

Generators with permanent magnets to produce the magnetic field suffer from two main disadvantages: permanent magnets are not very powerful compared with *electro magnets* made by passing current through a coil surrounding an iron core, and cannot therefore get the best performance out of a generator. Also there is no way of changing the magnetic flux of a permanent magnet, and it is often useful to be able to change the magnetism in a generator to control the power produced. Electro magnets can be made to give almost any magnetic flux from zero to their maximum by altering the power supply to the coil that surrounds them.

For these reasons most generators today have electromagnetic poles instead of permanent magnets. The poles are made of shaped pieces of soft iron surrounded by coils of wire carrying a dc supply. Generators differ mainly in the way in which this dc supply is arranged. Direct-current generators can usually supply their own *field current* supply, as it is called, by diverting a small amount of the dc they produce into the *field windings* which form the electromagnetic poles—they can start themselves because there is usually a small amount of permanent magnetism (*residual field*) left in the pole-pieces, and this is enough to produce a small amount of dc when the generator is started up. Once this has been fed into the field coils the magnetism builds up rapidly and therefore so does the power output. Later in this chapter, when considering the renovation of secondhand generators, we shall see how to deal with a dc generator where the residual field is too low to start power production.

Generators that produce ac (*alternators*) do not allow such a simple solution to the problem of supplying a dc field current. Some large generators have what is effectively a small dc generator mounted on the same shaft as the large ac rotor, and some small ones have a dc supply from a storage battery that is charged up every now and then; but most small to medium-sized ac generators take their dc field current, or exciting current, by diverting a small amount of the ac output to a *rectifier* that converts ac to dc. In car alternators, where the output is designed to charge a storage battery and to run equipment meant for dc operation, the whole output is rectified to dc, and some of this is easily taken for the field current.

When alternators are used in this way, it is often convenient to use an adaptation of the revolving field type: the stator, which then supplies the output, can be quite a large coil, while the electromagnets, which need only a small amount of dc current, can be fed from slip rings without much load on the brushes or contacts.

It may seem odd to use an alternator in a car, where a dc supply is necessary, but the alternator has several advantages over the dc generator.

1. There is a definite, rather low limit to the running speed of dc generators, as at high speeds sparking occurs between sections of the commutator, and also it is difficult to keep the brushes closely aligned to the commutator surface, particularly after some wear has occurred. This poor contact itself leads to overheating of the brush/commutator contact points, and the commutator segments may become melted or pitted, which makes matters worse.

The alternator needs only two simple slip rings, which do not incur much wear and therefore give a smoother and more efficient contact.

2. Because they can be run faster, alternators can be smaller and lighter than dc generators in the same wattage range. For example, a normal 472W dc generator for a car weighs 26½lb (12kg), and measures 8in × 5in (203mm × 127mm). A 607W alternator (the nearest equivalent size) weighs 18lb (8kg) and measures 6in × 6in (152mm × 152mm).

3. Alternator maintenance is less, because the commutator and brush gear give most of the trouble with a dc generator, and these do not exist in an alternator.

4. A dc generator needs a cut-out system to make sure that, when the engine speed of the car drops below the point where the generator is supplying 12V, the battery does not feed current into the generator. The rectifier system on the alternator allows only a one-way progress of current.

The only complication of the alternator used for dc production is the need for rectifiers, with the possibility that these may break down. In practice, however, they are remarkably robust as long as large currents are not run through them in the wrong direction. The rectifiers carry quite a large current, and need to be kept cool, usually arranged by mounting them on the end of the generator near the fan which also cools the coils: being thus mounted on a large, thick plate of metal also helps to dissipate heat—it acts as a *heat-sink*.

Using alternators

If you want a cheap generator, look around the scrap yards for a car which has been badly damaged in some area away from the engine, and get the alternator together with its control box. You can identify your alternator by its markings, if still legible, or by looking up the handbook for the particular car model. Some typical alternator details are as follows:

Lucas 11AC: a 3-phase alternator rated at a nominal 45A at 12V, normally giving about 43A when up to speed at 13.5V, or a working maximum of about 480W. It can be run up to 12,500 rpm without damage. The control unit of the *11AC* is a transistorized system with no moving parts, designed to equalise the output by limiting the field current when the alternator is running fast. Older models such as the *Lucas A2C* have a vibrating-contact control unit.

These control units are very useful for stabilising the output and making sure that the alternator itself is not overloaded if it is run very fast: all that happens with either type of control is that the field magnets get less current as the output rises, so the output goes back to a steady level.

Chrysler AC: a 12-pole alternator rated at a nominal 33A at 15V when the engine speed of the car is 2,200rpm, or about 500W output. It has a similar control unit to the Lucas model.

CAV 24-volt: this is typical of the larger alternators fitted to heavy transport vehicles, and a very useful source of power if you can get hold of one. A typical CAV model gives a self-limiting output of 60A at 27.5V, or 1650W, from a 12-pole alternator with rotating-field system. It can operate up to 4,500rpm (alternator speed, not engine speed), weighs 37½lb (17kg), and has a separate rectifier unit weighing 20lb (9kg) and a control unit weighing 6½lb (3kg). These weights are mentioned because they would be a serious consideration if the alternator were to be used, for example, in a wind generator mounted at the top of a mast.

There are also slightly smaller CAV 24V alternators with built-in rectifiers: these are the *AC5* range.

If you have acquired a secondhand alternator, try to get hold of the circuit diagram of the car from which it came, or note the connections if you are removing it yourself. Take special note whether the car is negative-earth or positive-earth.

The alternator will probably have four leads. One of these goes straight to the live side of the battery (ie positive for a negative-earth system and vice versa) and is the main supply cable. Another will go to a relay, probably via the ignition light, to illuminate the warning-light if your generator is not operating properly. The two others are the contacts to the field coils: one will go to the alternator control box directly, the other through the relay. You can connect the field-coil connection which runs from the alternator to the relay direct to the corresponding field-coil connection in the control unit, in most cases.

Connect up the alternator, alternator control unit, and a suitable 12V battery earthed to a water pipe or some similar earth line. Earth the earth lead of the control unit. If you have an ammeter (such as is used in many cars) capable of reading up to about thirty amps, put this in the circuit between the alternator and the battery. If not, fit a car lamp-bulb between the alternator and the battery.

Spin the alternator by fitting a buffing pad to an electric drill and applying it to the alternator shaft. If it is in good condition you should see action from the ammeter or the light.

Servicing of alternators, because of the variety of models, is rather beyond the scope of this book. However, the common faults are as follows:

1. *Slip rings dirty or corroded* Open up the end of the alternator nearest the rectifier diodes (there will usually be six of these at one end) taking care not to damage the connections to the diodes or to bend their pins. The slip rings and brushes should be fairly easily accessible. If the brushes are worn, replace them, and clean the slip rings with a clean cloth dipped

in petrol or paraffin. Do not try to machine the slip rings smooth: if they are really battered it is better to replace the rotor.

2. *Field windings faulty* In most alternators the resistance of the field windings, as measured between the slip rings, should be about 3–4Ω. If you do not possess a low-reading ohmmeter pass a current from a 12V battery between the slip rings with an ammeter in series. The reading should be 3–4A.

3. *Field windings shorted* Test for continuity between a slip ring and one of the rotor poles. If there is continuity, the insulation has broken down and the rotor should be replaced.

4. *Breaks in the stator windings* Check for continuity between the various lead-in wires. If there is no continuity there is a break somewhere in the windings. It would probably be cheaper to acquire another alternator and hope for better luck, if this is the case.

5. *Diodes faulty* Disconnect the diodes from the stator leads, using a pair of long-nosed pliers as a heat-sink when unsoldering. Check each diode with an ohmmeter: the resistance should be much greater in one direction than in the other. If both are low, the diode should be replaced. I do not recommend this test, or the unsoldering of the diodes, unless you have some previous experience with electronic components: you can easily damage a good diode by faulty work with the soldering iron.

Assuming that you have a satisfactory alternator, you can connect it to whatever driving system you have—engine, windmill, and so on. The electrical connections should be

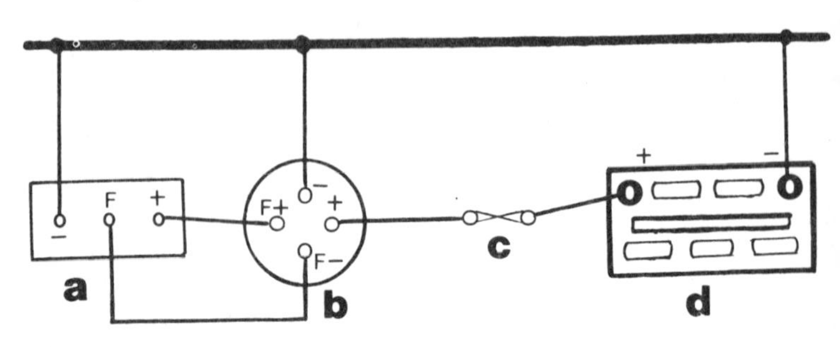

4 Connecting up an alternator. (a) Control unit, (b) alternator, (c) fuse, (d) 12V battery. Reverse the connections for a positive-earth system

as in Fig 4. (I have assumed a negative-earth system, which is normal in cars that have alternators: if you have components from a positive-earth system, reverse all the polarities in the connections.) As the case of the alternator and some other metal parts that, in the car, would be connected to the chassis, are meant to be negative, it is best if you keep this system throughout your electrical layout, connecting the metal frames or cases of equipment with the battery negative and the 'live' sides with positive.

As alternators produce ac as their initial product, it does not matter which way your alternator rotates. Be careful to fit a fuse as shown: apart from general safety considerations, the diodes will be ruined if there is a short circuit from the live side to earth.

It might be thought that it would be possible to get an ac supply from the alternator simply by cutting out the diode rectifiers. This would have the advantage that the ac output at about 13.5V could be stepped up to 240V by a transformer and used to run mains equipment. But there are two difficulties about this, even though the modification is possible, with a certain amount of electronic adaptation. The primary difficulty is that the field coils must have dc supply, and this means leaving the diodes in place and tapping off the ac supply from the ends of the stator windings—quite a tricky operation. The other difficulty is that the control unit which adjusts the field current to avoid overload also needs dc. If you are experienced enough to carry out these modifications, you will probably not be reading this section of the book. What you can do is to build a simple alternator with less complex problems (*see pp 38–40*).

Using dc generators

Many older cars found in dumps or being broken for spares, have dynamos (dc generators), and these can also be used successfully to provide an alternative source of power.

Normal car dynamos have two poles for the smaller models and four poles for larger, with the coils carried on the rotor and the magnetic poles on the stator. Current is collected from a commutator by two or four brushes. The field current is supplied by taking off a certain amount of dc from the brushes (*shunt winding*), and, as with the alternator, there is a control box that adjusts the amount of current actually passing through the field windings, in this case by introducing a variable resistance into the field circuit, which increases as the output increases, thus reducing the current to the field coils and bringing the dynamo back to a steady output.

An essential piece of equipment that goes with any dynamo charging a battery is the cut-out. You should always secure this as well as the control unit if you get your dynamo by stripping it out of an old car. The cut-out is designed to isolate the dynamo from the battery if the dynamo speed falls so much that it is no longer delivering 12V or more. If there were a direct connection and this slowing down occurred, current would flow from the battery to the dynamo, running down the battery and, much more important, letting the battery current, which runs in the opposite direction to the normal dynamo output, find its way into the field windings. Having this reversed current passing through the field coils would reverse the polarity of the magnet pole-pieces, and when the dynamo got up to speed again the current it produced would be also reversed. Thus, instead of charging the battery, it would put all its force behind discharging it, and both the battery and the dynamo would probably be ruined by the heavy current loss. If your system has a storage battery in it, you must therefore install the cut-out and make sure it is working efficiently.

Dynamos do not usually have the same output as alternators, and the 12V car type usually give from 2–400W. Because of the difficulties with the commutator, already mentioned, top running speeds are about 9–10,000rpm. The heavy-duty 24V dynamos in use on some commercial vehicles, such as the *Bosch LJ/GTL* range, can provide about 1,000W, but unless you happen to see one of these going cheaply from a breaker's yard you would do better to use a 24V alternator such as the CAV type, which is more powerful and lighter.

Using a dynamo

If you have acquired a secondhand dynamo, control unit and cut-out, assemble them with a battery, taking care to get the polarities right (most dynamo-driven cars of the older sort tended to have positive-earth systems). Again, as with the alternator, a circuit diagram from the car handbook will help, or if you cannot lay hands on this mark the connections before you disconnect the equipment from the car. It is not vital to use the voltage regulator when testing the dynamo, but be sure to incorporate the cut-out in the circuit. Put a lamp or ammeter in circuit between the cut-out and the battery, and spin the dynamo with an electric drill. (*See p 31*). You should see signs of a current passing opposed to the battery current.

To service a dynamo, look first at the commutator, which is the usual source of trouble. You may find that the brushes are sparking or arcing, or just not connecting properly. Fit new brushes if the old ones are worn, and check that the brush springs are giving the right pressure—this is more important with a commutator generator than with one fitted with simple slip rings.

Now look at the commutator. It will probably be dirty, with some of the copper contacts worn and scratched, and some or all of the insulating pieces between the copper contacts sticking up. This is because the insulators wear at a different rate from the copper, and the effect is to make a series of bumps on the commutator over which the brushes bounce, thus creating sparks and bad contact. (*See Fig 5*).

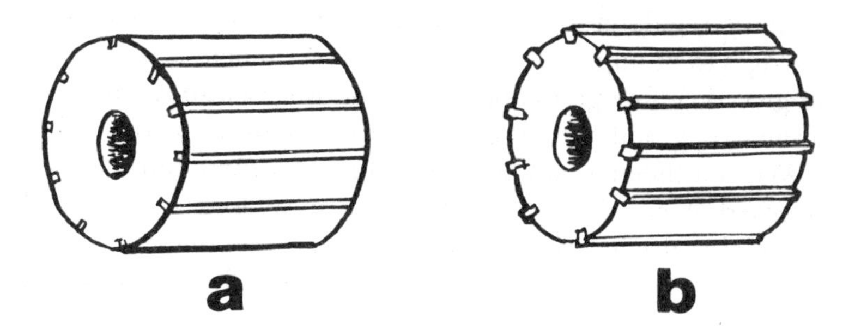

5 Commutators: (a) in good condition, (b) in bad condition. The effects of wear have been exaggerated for the drawing

If you have access to a lathe, fit the commutator between centres and skim down the whole surface to eliminate scratches and so on. If not, fit a power drill to the shaft of the armature by a flexible drive, rest the armature on two suitable bearings, and spin it while holding very fine emery paper against the commutator. This should smooth the copper, but may leave the insulating spacers looking rather ragged. Go over each insulator with a fine hacksaw blade held in line with the shaft of the armature, and very gently rub down the insulator until it is smooth and level with the copper surface. Do not cut down below the copper level if you can help it, but do not leave insulation material sticking up.

Other faults are likely to be in the field windings, in the stator. You will find the wires from these coming out and connecting up with the control unit. (Another wire from the dynamo to the unit carries the output current which is 'rationed' to the field coils by the unit, but it should be fairly easy to tell which is which, even without the circuit diagram.) The resistance of the field coils should be about 4–6Ω, depending on the actual model. If it is much less there is a short in the windings, and the coil will take up too much current; if it is very high there is some break in the windings and the dynamo will not give a satisfactory output.

If the cut-out has been badly adjusted, you may find that it stays open-circuit (ie it does not allow current to flow) even though the dynamo is producing more than about 13V. You will find an adjusting screw next to the cut-out contacts, in most regulator boxes, that allows the cut-in and cut-out voltages to be adjusted. Do not turn it very much: it is designed to give very fine gradation of control. Most cut-outs cut in at around 13V and cut out when the voltage drops to 10V.

When a dynamo has been left unused for some time (as in a wrecked car), the field pole-pieces may lose the residual magnetism needed to make the dynamo start producing current. You can revive the magnetism by giving the field coils a short burst of power from a battery. Assuming that your system is positive-earth, fix a battery, a 12V bulb and the dynamo in series as shown (*Fig 6*). The bulb is to limit the flow of current through the field coils. Only let the current flow for about five minutes; the idea is to give the pole-pieces of the field magnets just enough magnetism for them to start current-generation as the armature moves. As soon as a small amount of current is circulating, the portion of it diverted back to the field coils will bring the magnetism, and therefore the current, up to full strength very rapidly.

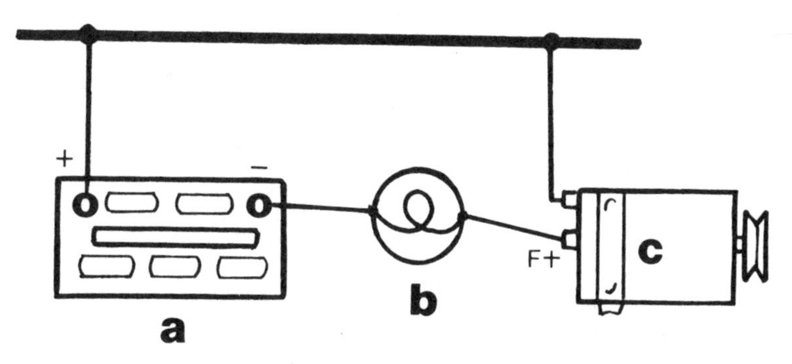

6 Reviving the field magnets of a dynamo: (a) 12V battery, (b) 40W lamp (12 volt working), (c) dynamo. Reverse the connections for a negative-earth system

Rewiring a dynamo

If you have a scrap dynamo with bearings and mechanical parts in good order, but with some permanent electrical fault that makes it unusable, you may like to consider rewiring it as an alternator. It could be used to provide ac in a suitable state for stepping-up to 240V through a transformer, as long as you have some source of dc, such as a battery, to keep the field coils fed.

First take the dynamo apart. Take off the pulley from one end of the shaft (if the dynamo has been lying around for a long time you may have to use a fairly powerful puller to do this without damaging the shaft) and unscrew the long bolts that run between the two bearing plates. You can now slide the bearing plates off the shaft of the armature. Take out the paper fillers stuffed into each slot of the armature, and cut the wires at the commutator. You can now remove the windings and clean the armature. Note the dimensions and position of the commutator, as your slip rings will have to be placed in the same position and connect with the same brushes. Now take off the commutator and clean the shaft.

To construct a pair of slip rings, which must be fitted instead of the commutator, find a piece of copper pipe of the same external diameter as the old commutator, or very slightly larger, and carefully saw two rings from this with a hacksaw. Use a blade with fine teeth, lubricate it often, and try not to press too hard and therefore distort the rings. Saw off a piece of pipe slightly less than half the commutator length, so that when you put the rings together with a small gap in between the slip ring assembly occupies the same volume as the commutator you have removed. Rub the rings smooth with fine emery paper to remove burrs from the saw.

Find a Tufnol or similar insulating bush which can slide over the shaft and has an external diameter about equal to the internal diameter of the slip rings. You may have to make one to fit: if so, it is easier to find one of the right diameter and enlarge the central hole by drilling. Be careful to keep the hole central so that the bush will run smoothly without any eccentric

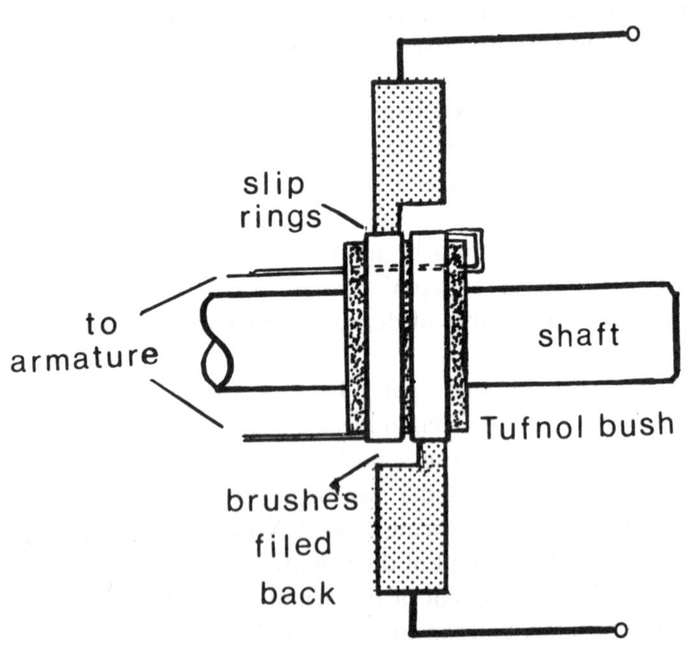

7 Slip ring system for converting dynamo to alternator

motion (*see Fig 7*). Mount the bush on the shaft where the commutator was, using epoxy-resin adhesive (Araldite), and stick the slip rings side by side on the bush with the same adhesive.

File away part of the rubbing surface of each carbon brush (*see Fig 7*) so that instead of covering the whole area it only contacts one slip ring. Drill a fine hole through the bush at one point so as to take the wire through from the windings to the outer slip ring.

Now the armature has to be rewound. First insulate the iron laminations, either by painting the whole armature core with a thick layer of polyester resin (preferably the slightly elastic type, which will stand up better to vibration), or by winding the whole core with a layer of cloth insulating tape. Now wind the slots with twenty-four turns of 18swg enamelled

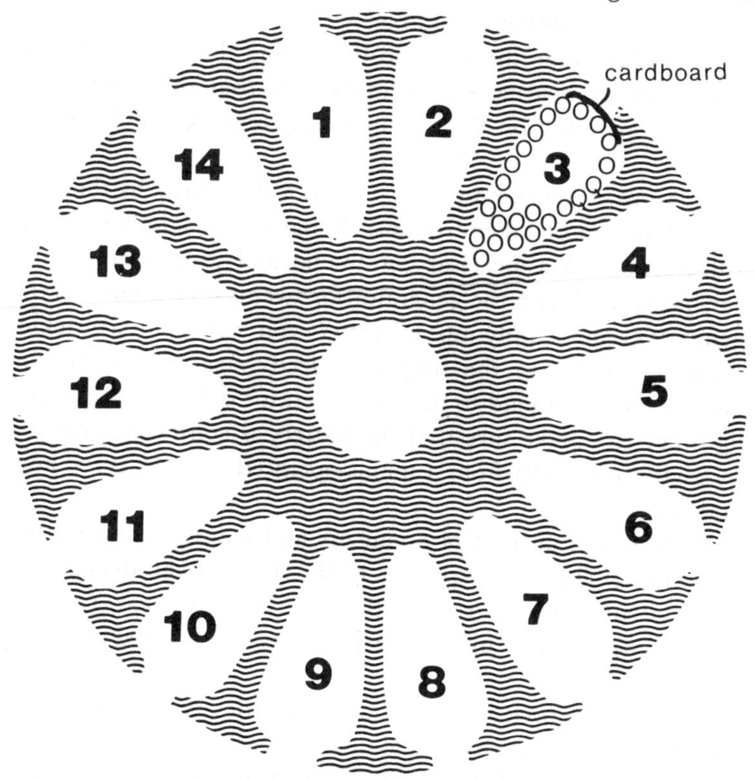

8 Rewinding a 14-slot armature. Winding 24 turns of wire in each coil, proceed as follows: wind a coil between slot 3 and slot 14, then between slot 7 and slot 10, then between slot 4 and slot 13, then between slot 6 and slot 11, then slot 5 and slot 12, and bring out the wire. Connect the start to the outer slip ring (furthest from the armature), and the end to the inner slip ring

wire in each pair of slots: Fig 8 shows the order of winding for the common 14-slot dynamo armature. Start with the slot which is in line with the hole you have drilled through the plastic bush to carry the wire to the outer slip-ring, and go on until you have completed the winding, taking the end of the wire to the inner slip ring. Solder the ends of the wire to the slip rings, leaving no bumps of solder on the surface to be rubbed by the brushes, tie the ends down with thin string round the armature, and cut small pieces of card to fit over the wiring in each slot. Now soak the whole assembly in thin shellac solution—the liquid sold as 'knotting' for covering knots in wood before painting is suitable.

Now reassemble the dynamo by putting the armature shaft back into the bearing plates, make sure that the brushes are running smoothly on their proper slip rings, and bolt up the case.

To work, the generator must have a dc supply to the field coils. This can be taken from a 12V battery: a supply of about 1.8–2.0A at 12V is necessary, so the normal 50Ah (amp-hours) car battery will run the generator for twenty-five hours. As the generator can supply up to 150W you could fit a trickle-charging device to the output to feed back current to the battery. If you intend to use the generator only for emergency lighting, etc, make sure that the battery is constantly charged by giving it a boost from a mains-operated charger every now and then.

With a field current of about 2A and run at 3,000rpm the generator will give 35–40V at 50Hz (the voltage depends to some extent on the load which the generator supplies). You can step this up to 240V by using a 1:7 transformer rated at about 200–250W.

It is a good precaution to have some device to cut off the field current if the driving force fails or the generator stops working for any other reason. Apart from the drain on the battery, the field coils will get hot if the armature is not turning, and could damage the surroundings. A simple means of avoiding this is to take a line back from the high-voltage side of the transformer, and use this to operate a 240V relay (contacts *open* when not energised), between the battery and the field windings. As long as the generator is supplying a suitable voltage to give 240V on the high voltage side of the transformer, the relay contacts will be closed and the battery connected to the field coils, but as soon as the voltage falls below about 180V due to any faults, the relay contacts will open and cut off the connection to the battery. A press-button switch must also be

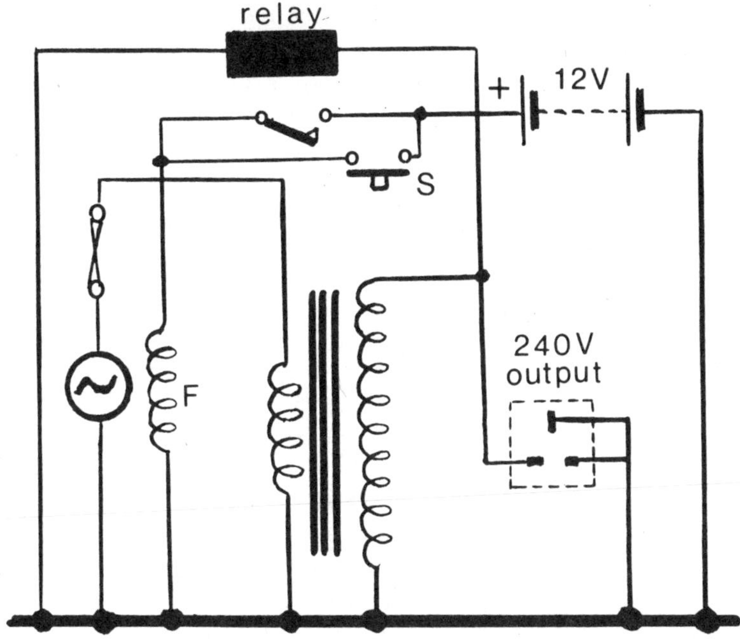

9 Circuit for alternator giving 240V ac. Switch S is a press-button type which opens again when released. The relay has a 240V ac coil and contacts rated for 5A (contacts open when not energised)

connected into the circuit to start the field current (*see Fig 9*): pressure on this switch will connect the battery direct to the field coils, and, if the alternator is running, there will be an immediate production of 240V current that will operate the relay and close its contacts. If the finger is now removed from the press-button the generator will go on operating even though the direct line from battery to field coils has been cut off.

The last stage is to connect the output to a 3-pin socket, as shown in Fig 9: connect the N and E contacts of the socket together to the earth line of the circuit. This socket will then supply any piece of equipment up to about 150W.

The frequency, as has been said, depends on the speed of the drive. If you have good speed regulation you can get a fairly steady frequency at 50Hz, but if your motive power is variable it is obvious that the frequency will also vary. This does not matter at all for ordinary filament lamps and similar equipment, but it may upset fluorescent lamps, which are made to run most efficiently at a fixed frequency. Electronic equipment such as radio and TV is not affected too much by a small variation of frequency, because most of the ac supplied to such sets is rectified to dc before use.

Apparatus driven by motors of the synchronous type suffers worst from variations in frequency. These motors depend on the regular pulse of the alternating current to keep them turning, and the speed is directly dependent on the frequency of the supply. For example, we all know that when the electricity authorities shed some of their load by lowering the frequency (a common method of dealing with a potential overload during the winter), electric clocks begin to lose time, because their synchronous motors are running too slowly. The clocks have to be corrected by running the frequency a little higher than 50Hz after such a load-shedding. The simple home-made generator described here is not likely to give very good results with synchronous motors unless you can fit the drive mechanism with some fairly sophisticated control gear (and in that case you are unlikely to build such an elementary generator anyway).

On the other hand, it must be said in favour of such alternators that the output is almost sinusoidal, and very much more like mains supply than the output of most inverters (*see page 100 for details of these*). Most inverters tend to produce *square-wave* outputs, where the alternations are cut off sharply and abruptly, not smoothly. This tends to waste power, especially with induction motors, and on an ordinary voltmeter will tend to give lower readings than the true average voltage. There is a danger, therefore, when using a power supply with square-wave characteristics, of putting too high a voltage into such equipment as TV sets, which need a carefully regulated voltage supply for the filaments of the picture tubes and any other valves they may contain. This aspect of power supplies is considered more fully in Chapter 10, which deals with the production of 240V ac from low-voltage dc supplies.

The alternator described above gives a smooth ac output that will give accurate rms voltage readings on an ordinary voltmeter. If the voltmeter says 240V, set the voltage tapping of the set to this figure; if 220V, to the lower setting, and so on. Make sure that you carry out the voltage readings with the set connected momentarily. If you just measure the output voltage with no load, you will find that it goes down as soon as a reasonable load is connected.

Other generators

The simple generators described up to now, mainly cannibalised from car parts, have been intended for the reader with not much money to spare, but wanting to supply at least part of his electricity himself. None of the generators in this class, however, will give much more than 600W at best, hardly enough for a household supply, even allowing for storage during the night and slack periods of use.

By using an alternator from a heavy commercial vehicle you can get up to about 1200W, but these generators are not as readily available as the car type, and are likely to cost you quite a lot, even secondhand, especially if you buy them as reconditioned and guaranteed to work.

If you want more output than this, the only solution is to buy a commercially made generator. Quite often dealers in secondhand generating sets have separate generators for sale, because the petrol or diesel engines of these sets tend to wear out before the generators themselves. Alternatively, look around the trade papers; many factories use dc generators to supply dc motors and similar equipment, and plating or welding shops may have very heavy-duty models; you may be able to pick up a generator from a factory that is re-equipping or being closed down. Look for auctions of industrial equipment: while the dealers are haggling over the fork-lift trucks you may be able to sneak in a moderate bid for a generator. However, do not expect to get one all that cheaply: if a large generator is very cheap it is probably burnt out.

One generator deserves mention because it is specially designed for driving by wind power. A difficulty with wind generators is that the vanes or propellers tend to turn rather slowly compared with the ideal speed for the generator (*see Chapter 6*). Gears or belts add to the cost of the installation and also consume some of the power, and if the generator is geared up to give a reasonable speed of rotation in average winds, it may be severely overrun in high winds or gales, with damage to the generator or the equipment it is supplying.

Joseph Lucas Ltd have a specially made alternator, the 17 ACR model, which is intended for use with wind-vanes or propellers. It can start producing useful electricity at as low a speed as 450rpm, but is rated to run up to 15,000rpm without damage, thus coping with the very wide range of wind-speeds encountered. At about 12mph it will produce 5–7A at a nominal 12V, which is rectified by built-in diodes to provide battery-charging facilities, and at slightly higher average wind-speeds it can produce a steady 150–180W.

5 Engine-driven generators

A generator driven by a petrol or diesel engine has obvious advantages, especially for emergency supplies. The main benefits are as follows:

1. The engine can be arranged to run at a constant speed, and therefore the output is also steady.
2. This constant-speed running also means that, by proper choice of generator, the supply can be provided at the required voltage and frequency without any intermediate apparatus such as transformers or inverters.
3. For emergency use, the fuel is easy to store and requires very little attention, unlike storage batteries that need a lot of attention and involve a high capital cost.

On the other hand, the engines are all fairly noisy, and at present prices the cost of providing electricity from a petrol or even diesel engine is higher than the cost of mains electricity. This limits use of these machines to emergency situations. The price of the public supply of electricity is going up all the time, but this is due principally to the rising price of petroleum, so the running costs of the generator will also rise.

Accepting these limitations, many people will be thinking about the purchase of an engine-driven generator as a precaution against power cuts.

Generating sets are available from a number of manufacturers (*see Appendix 4 for suppliers*) in a range roughly from 250W output (for example the Honda E 300 range) to 650kW (the Lewis LD 650, which is an industrial generator capable of supplying the domestic needs of a whole village). The smaller sets (250W–4kW) are usually fitted with petrol engines but diesel engines, which are cheaper to run, are available for all the generators of about 1.5kW and over.

There is a slight complication in the rating of many of the large generators, because power factor losses come into consideration. This means that the figure obtained by multiplying the voltage by the current in amps is rather higher than the actual wattage produced: a power factor of eighty per cent is common with the larger machines. Because every manufacturer likes to quote the highest figure for his machines, the ratings are usually given not as kW but as kVA (the actual V × A multiple), so that a 200kVA machine with eighty per cent power factor will actually only produce 160kW real power under most conditions. It is necessary therefore to choose a machine with a rather higher kVA rating than the load in watts to be supplied. Many manufacturers quote the power factor of their generating sets in their technical literature.

If you live in a very high spot, remember that petrol and diesel engines run less efficiently at high altitudes, because of the reduced atmospheric pressure. If you live at more than 500ft (152m) above sea level, deduct three per cent from the

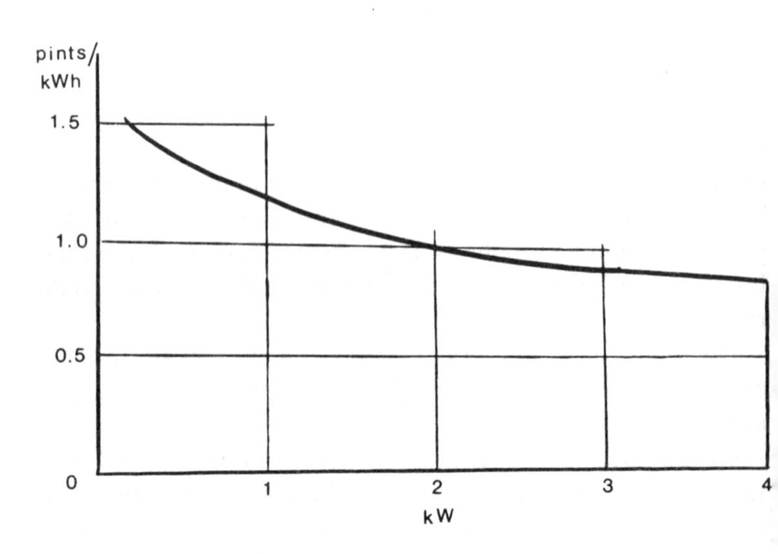

10 Average consumption of petrol-driven generating sets, pints of petrol per kWh for various sizes of unit

rated power output of the machine for every 1,000ft (304m) altitude over the 500ft level; so if you live at 2,500ft (760m), your 200kVA machine will really only produce about 150kW.

The cost of running a generating set diminishes, as one might expect, as the engine gets larger and greater efficiency is possible. Fig 10 shows approximate consumption of petrol per kWh for various sizes of generator from 250W to 4kW, and

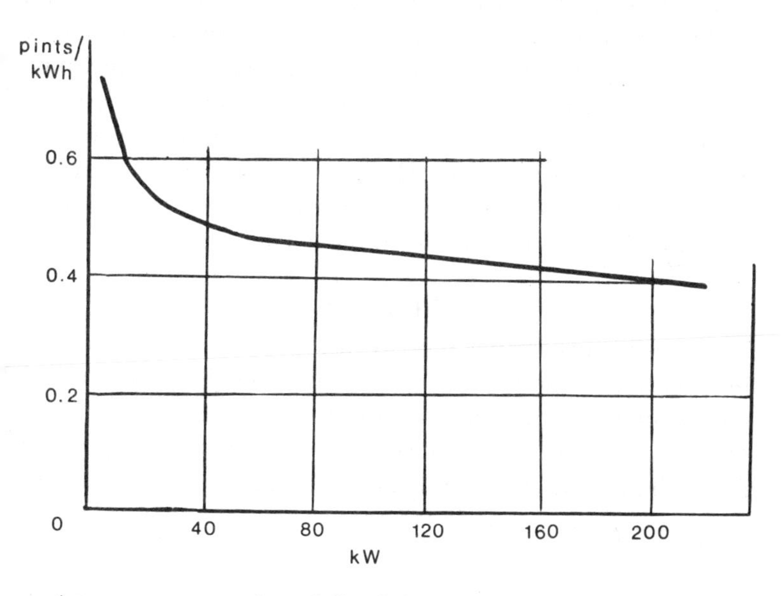

11 Average consumption of diesel-driven generating sets, pints of diesel fuel per kWh for various sizes of unit

Fig 11 shows similar figures for diesel generators from 1.6kW to 200kW. Obviously these figures will vary somewhat from one type of engine to another, but they should give a guide to the probable cost of electricity per unit from the generators.

Noise

Some of the smaller generators are quite well designed to eliminate much of the engine noise—the Honda EM 300, for example, produces about 58dB (decibels) at 5m, about the noise level of a quiet conversation, but some of the large sets are extremely noisy, not tolerable in the house.

The best solution is to house the generator in a separate shed or outbuilding, and pay careful attention to the sound insulation: rock wool or glass wool panels around the walls will help, and making sure that all spaces between the roof and walls are thoroughly plugged will also help; sheds with corrugated iron or asbestos roofs often have spaces where the corrugations do not meet the top of the wall. These holes must be plugged with some heavy filler, preferably sand and cement, otherwise a lot of noise will escape.

The exhaust of a generator engine causes problems, because it obviously cannot be shut in an enclosed building, yet a great deal of noise is transmitted from the end of the exhaust pipe. Some manufacturers supply baffles and silencers; it is worth inquiring about this if your generator is to be used near your own or someone else's house. Home-made silencers may interfere seriously with the running of the engine and thus produce unpleasant fumes and cut down the efficiency.

Starting

The smaller generating sets are usually started with a cord, like outboard motors and lawnmower engines, but this is inconvenient or even impracticable with larger units. Many of the larger sets have a self-starter of the car-engine type, run from a battery, and can easily be adapted to start automatically if the mains power supply fails.

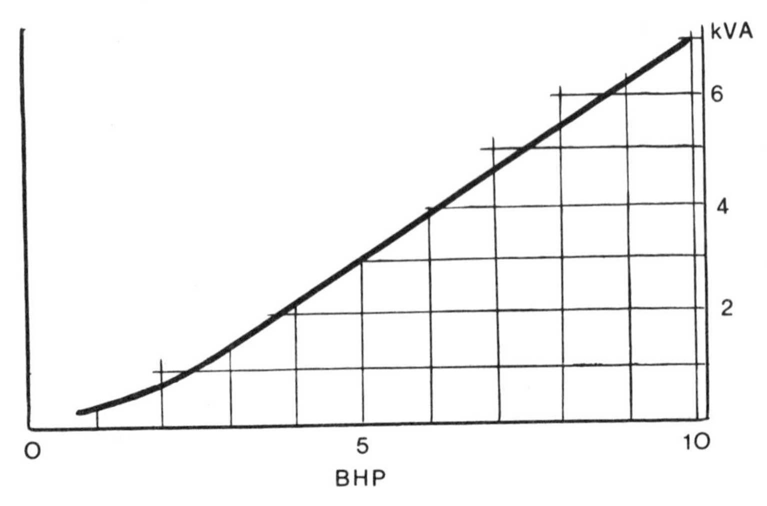

12 Power of engine necessary for driving generators of various capacities, BHP against kVA

Home-made generators

It is quite possible to run an alternator or similar generator from any engine you have available, if you want to have the advantages of an engine-driven system without the heavy capital outlay. Figure 12 shows an approximate guide to the brake horsepower you will need for various capacities of power generation. The small engines used in lawnmowers will do very well for generators up to about 300W, and for larger sizes motor-cycle engines or small car engines can be used. Such a set-up is unlikely to be economic in terms of cost of fuel per unit, compared with the cost of mains electricity, but for emergency use it has obvious advantages. If you leave the main shaft pulley on a small car engine, this provides an ideal point for a belt-drive to the generator; with a motor-cycle engine of the chain-drive type you could use a small sprocket on the generator shaft and keep the chain-drive.

Storage of fuel

If you have a petrol-driven generator, in particular, be careful about the storage of fuel for the engine. In fact the storage of more than two gallons of petrol is illegal in the UK unless special fire precautions are taken. If you have a large installation, consult the suppliers about the management of a fuel tank and supply system.

6 Wind generators

The wind is a vast untapped source of power, produced by a combination of solar heat, which causes thermal air movements, and the rotation of the earth, which keeps the trade winds and other steady breezes blowing in the opposite direction. One calculation has shown that the total wind energy at present wasted, over the whole world surface, is just about twice the output of all the power stations in existence. Wind power can be plentiful where solar heat is less reliable: while the sun often hides itself from Britain, Holland and Denmark, for instance, these are among the windiest of the world's developed countries.

Our ancestors, not yet seduced by the easy power of the steam engine, made ample use of the wind. Around 1750 there can scarcely have been a skyline in northern Europe without the familiar rotating cross of a windmill—10,000 in England, as many in Germany, 8,000 in Holland, and about 5,000 in Denmark. After the Industrial Revolution, the prudent Germans, Dutch and Danes kept their windmills going, alongside the steam engines. In 1895 Germany had more windmills than before the coming of steam (about 18,000), and the Danes, with little coal and no water power, still use windmills to grind corn, pump water and, more recently, to generate electricity by turning dynamos.

Official interest in wind power on a large scale has waxed and waned. Large wind generators have often been designed and built in a bright cloud of optimism, and then abandoned because development funds have been cut off. One of the most heroic generators was built in the USA in 1941. Standing on a windswept hill in central Vermont, Grandpa's Knob, this was a 110ft steel tower carrying a giant stainless-steel propeller

175ft across, driving a generator large enough to turn the power from a 30mph wind into 1.5 megawatts, enough to supply electricity to about 500 homes. Unfortunately this monster had to be taken down when the propeller cracked, and wartime conditions then prevented further development.

Small wind generators to supply some or all of the power for a single household can be made without specialised skill or materials. The few simple calculations that follow are intended to make it easier to decide on the type and dimensions of a wind generator for individual requirements.

Energy from the wind

The most important point to note is the effect of wind speed on the extractable power. The harder the wind blows the more energy it must produce, but in fact the speed is far more critical than might appear at first sight, because the energy in the wind is proportional to the cube of the speed: a 20mph wind contains eight times more power than a 10mph one, and a 30mph wind twenty-seven times as much, and so on. This makes it more important to choose a spot with the highest average wind speed, and it also means that designs have to cope with a very large difference in output.

(For the mathematically minded, it may be said that this cube relationship exists because, if the weight of a mass of air passing a certain point is m, and its speed V, the energy it contains is $\frac{1}{2}mV^2$. However, with wind, the weight of air passing the spot is also proportional to the speed, so the energy is proportional to V^3.)

The other factor that affects the amount of energy is obviously the area over which it is absorbed—a 100ft diameter vane will be able to take in more energy than a 10ft one. The total energy in a wind of speed V crossing an area A is found to be 0.00064 AV^3 kilowatts, if A is in square metres and V in metres per second, or 0.0000053 AV^3 kilowatts if A is in square feet and V in mph.

Unfortunately it is not possible to use all the energy that these formulae suggest. If you could design a windmill that extracted all the power from a moving mass of air like the wind, the air would stop moving as soon as it was past the windmill, and a vast mass of stagnant air would build up like a cushion, hampering further movement. Some energy must be left in the wind to maintain the air-flow and make sure that the mill keeps turning. The theoretical maximum extraction turns out to be 0.593 times the figures above, or 0.00038 AV^3 (m^2:m/s) and 0.0000031 AV^3 (ft^2:mph).

Added to this difficulty are the various losses in the windmill and generator from friction, electrical leakages, and so on. These vary with the design, but usually mean that only forty to seventy per cent of the theoretical extractable energy is actually converted into electricity for use. Assuming forty per cent efficiency the final figures for useable power come to 0.00019 AV^3 kW (m^2:m/s), 0.0000015 AV^3kW (ft^2:mph).

Table 5 gives an idea what these formulae mean in practice: the top line is average wind speeds in mph with the region of the Beaufort scale added as a practical guide, and the figures down the side are diameters of a windmill sail or propeller, in feet.

Table 5 Windmill output at various wind speeds

	Wind speed (mph)							
Windmill diameter	*Light breeze*		*Moderate wind*		*Strong wind*			*Gale*
(ft)	5	10	15	20	25	30	35	40
1					17W	30W	47W	70W
2	Under	10W	15W	35W	68W	119W	188W	282W
3			33W	79W	155W	267W	424W	634W
4		18W	59W	141W	275W	475W	755W	1.1kW
5		27W	93W	220W	430W	742W	1.2kW	1.8kW
8		70W	238W	563W	1.1kW	1.9kW	3.0kW	4.5kW
10	13W	110W	371W	880W	1.7kW	3.0kW	4.7kW	7.0kW
15	30W	247W	835W	2.0kW	3.9kW	6.7kW	10.6kW	15.8kW
20	55W	440W	1.5kW	3.5kW	6.9kW	11.9kW	18.9kW	28.2kW

It is sometimes difficult to assess wind speeds unless you have an anemometer: Table 6 gives some hints based on ordinary observations of the effects of wind.

Windmill designs

These can be classified roughly as horizontal-shaft mills, like the conventional corn-grinding mill, and vertical-shaft mills, which are rather like turnstiles driven by the wind. Of the horizontal-shaft mills there are three main types: *slat-type mills,*

Table 6 Simple assessments of wind speed

Wind speed (mph)	Observations
Under 1	Smoke rises vertically from chimneys
1–3	Smoke drifts slowly
4–7	Breeze felt on face. Leaves on trees rustle
8–12	Leaves and twigs in constant motion
13–18	Dust clouds raised. Small branches sway
19–24	Small trees in leaf begin to sway
25–31	Large branches in motion. Audible whistling in telephone and other overhead wires
32–38	Large trees sway. Walking against wind difficult. Wind howls round buildings
39–46	Twigs break off from trees. Difficult to stand straight in wind

with framework sails similar to those used in the traditional corn-grinding windmill; *multi-blade fans* or windwheels; and *propellers,* much like those used on small aircraft, with two, three or more narrow blades.

Slat-type mills

These have not been in much favour for electrical generating plant, partly because the sails require skilled carpentry, partly because they are slow-moving, and quite a lot, I suspect, because designers wanted their generators to look 'modern'.

The first point need not trouble the home constructor making a small mill: while constructing a traditional 60ft diameter sail would need a steady eye and a lot of experience, a 6ft sail can be made by any competent handyman using fairly cheap materials.

The second point, the speed of rotation, is more serious. As we saw in Chapter 4, the output of a generator depends basically on the number of coils of wire passing a certain point in a fixed time. If the coils move slowly there have to be more of them to produce the power, so the generator has to be larger, and heavier. The only alternative, if we have a slow-moving source of power, is to run it into a gearbox that will step up the shaft speed, and this means extra cost and some loss of power by friction in the gearbox.

This has been the main argument against the slat-type mill and for the propeller, which runs much faster for any given wind speed and therefore often needs no gearing. However, there are arguments on the other side. So far, in the relatively short history of wind-powered electricity generation, all the big propeller-driven mills except one have failed because of mechanical faults in the propellers themselves, usually cracking or even disintegrating at high rates of rotation. Even with modern resources and engineering and materials, it is difficult to design a propeller that will run smoothly in all wind conditions. On the other hand, the slat-type windmills used for brinding corn and pumping water have often run for centuries with only the miller himself and the village carpenter to maintain them; so perhaps a slow-running mill has the attraction of reliability. The slat-type mill has great attractions also for the home constructor with simple woodworking tools: with patience and care, its making will give no trouble.

There are many traditional designs for mill sails, but one of the most efficient ever developed was the warped sail. This is set at an angle to the wind, but the angle varies from the centre to the tip of the sail. Figure 13 shows the accepted angles (or 'weather') for a warped sail. The angle is sharper at the

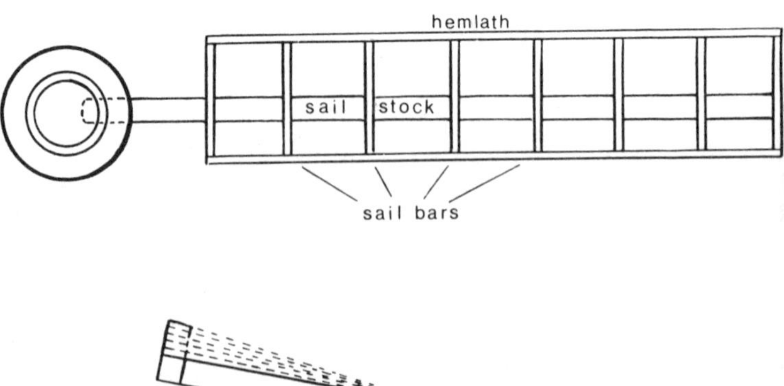

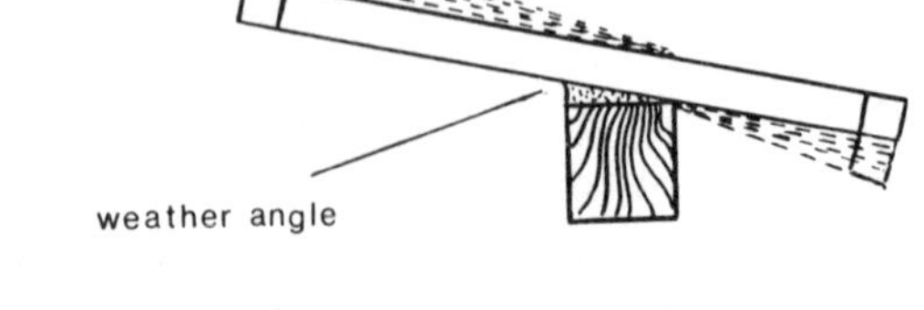

13 A warped sail. The lower drawing shows the view from the outer end of the sail, and the gradual increase in weather angle towards the centre. The outer sail-bar should be at 7° to the sail-stock, the inner one 18°

centre than at the tip: this is to allow for the greater air resistance met by the tip of the sail, which of course travels faster than the inner sections as the sail goes round.

Figure 13 shows the dimensions for a single 3ft sail made up in wood. Four of these sails, forming a 6ft mill, would be ideal for driving a 5–600W car alternator. The strongest part must obviously be the main shaft, the sail-stock, and if possible this should be made in a tough hardwood such as ash or hickory about 1½in (38mm) square, or in rather thicker softwood. This may seem heavy, but in a high wind a windmill meets powerful forces. You can sometimes find suitable pieces of hardwood in the framing of old furniture.

The framework of spars at right angles to the sail-stock, the sail-bars, is made up of 8in (203mm) lengths of 1in (25mm) square softwood. To get the correct angle at which each crosses the sail-stock, use packing pieces of thin wood or even cardboard under the raised edge before you nail the sail-bars in position, and then fill the triangular space underneath with putty or waterproof filler. You can make a good filler by taking a thick paste of cellulose filler, such as Polyfilla and water, and mixing this with linseed oil or old paint. When it sets it is water-resistant. Note that the sail-bars are not set symmetrically on the sail-stock, but cross it about a third of the way along their length.

Finish off the framework by fixing the edge, the hemlath, with oval nails, on the ends of the sail-bars.

In large traditional mills, the four sail-stocks are fastened together in a large central boss called the poll-head: this was mainly because it was almost impossible to find timber in straight pieces 60ft long to form two sails. It would be easy to find a 6ft length to form two of the sails for our small windmill, however, and to mortice the other two 3ft sails into the centre of the long piece. I suggest adding two iron straps or an iron cross-piece to reinforce the mortice joints, as any shrinkage of the wood as it weathers could allow the two short pieces of sail to come away from their mortices and perhaps do some damage in a high wind. Alternatively you can make all the sails the same length and mortice them into a poll-head, as if putting spokes into the hub of a cart-wheel: again an iron strap is a good precaution. This may offend the purist joiners, but even the old millwrights did not hesitate to use iron reinforcements at points of stress.

A covering over the framework of the sails collects the wind pressure. In early mills the frame was covered with canvas, and the miller would be kept busy taking in and spreading out canvas as the wind strengthened or died away, to maintain a constant sail speed. Later, shutters were fitted, flaps that opened when the wind was strong to let through some of the pressure.

For a small mill you may decide to have a fixed area covered. Fix strips of plywood on each sail-bar so as to cover all but an

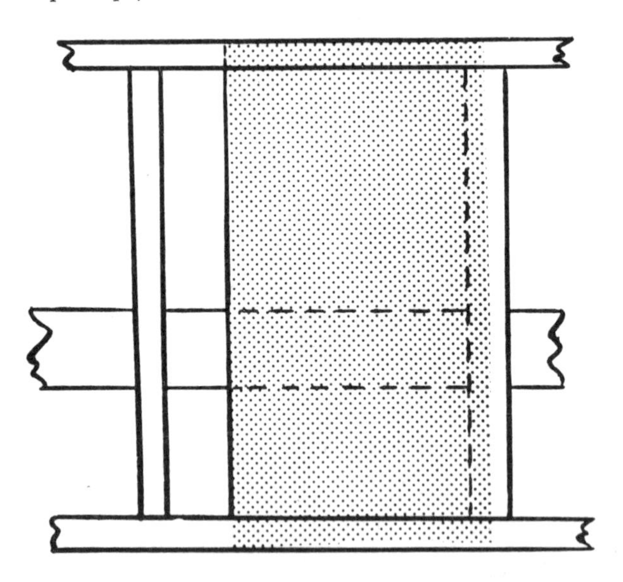

14 A simple slat for a sail, made with hardboard almost covering the gap between the sail-bars

inch of the space between sail-bars (*Fig 14*). These gaps will allow some wind to escape through the sail and make the action smoother. Alternatively these strips can be hinged to each sail-bar and held in place by a light twisted-wire spring, such as is used to hold a letter-box flap closed against the wind. This is obviously more trouble, but will make the mill far more resistant to sudden gusts of wind.

The shaft of the mill should preferably be steel rod, mounted in a ball-race. Fix a disc of steel perpendicularly at the end and drill it to take bolts, so as to fix it to the centre of the sail-stocks. The balance of the mill will depend on this mounting, so make sure that the shaft is really central. Minor differences of balance can be made up by fixing a small piece of lead to any sail that seems to be light, or by boring one or two holes and fixing nuts and bolts through the wood, until the complete mill-sail can be spun round on its axis without always coming round to the same point. The older millwrights used to fix their shafts so that the weight of the sails was thrown backwards towards the

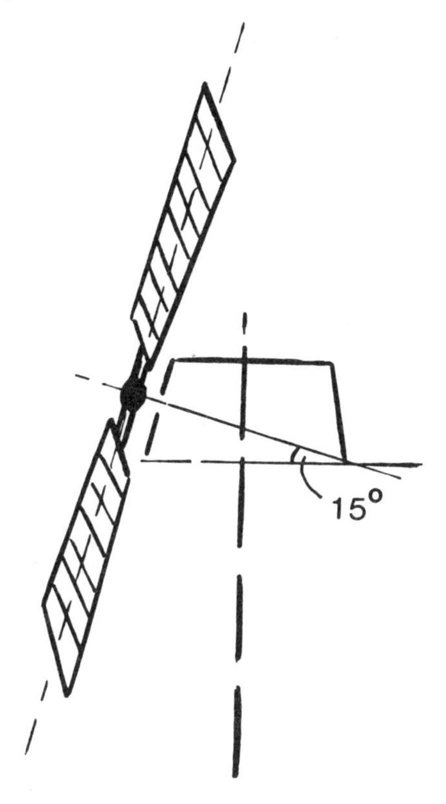

15 The shaft of the mill should be set back as shown at 15° to the horizontal

mill structure (*see Fig 15*): an angle of 15° to the horizontal was usual, and saved a lot of strain on the shafting.

Once your sails are mounted, you have only to arrange a drive to the alternator. A drive ratio of about 20:1 would be appropriate for a 6ft mill running a normal car alternator, and you can arrange this either with gears or belt drives; in either case, a gear wheel or pulley wheel will have to be fixed to the metal shaft of the sails with a key or grub screw, before it is finally set up. Whichever system you choose—gearing or belts and pulleys—make sure that it can safely transmit about 1HP. If you use car fan-belts and pulleys (an attractive method because the parts can be had for almost nothing from scrap dealers and your alternator will already have a drive pulley of the right size), you should have no trouble, but if you have a suitable set of gear wheels, this will do equally well.

Yawing affects all the horizontal-shaft mills, but may be conveniently considered here. A windmill of this type must always be facing into the wind, otherwise it will not work efficiently. So firstly the mounting of the mill has to be constructed so that it can turn freely to face the wind, or yaw, and secondly the electrical connections between the generator and the batteries or other load have to allow the mill to turn without tangling up cables or otherwise interfering with the electricity supply.

In the old full-size windmills, either the whole mill moved around an enormous central pivot (the post-mill), or a moving cap carried the sails, resting on a fixed tower (the smock-mill). For a smaller mill such as this, it is satisfactory to mount the mill and its shaft on a circular piece of ¾in (19mm) plywood or chipboard cut into a 15in (380mm) circle. This is bolted to a scrap car-hub (complete with the roller bearings—a front-wheel hub from a rear-wheel-drive car is easiest to remove) in the wheel position, and the inner (fixed) part of the hub is then bolted to a similar circle of plywood or chipboard fastened to the fixed base. (*See Fig 16.*)

Electrical connections can be dealt with in two ways. Unless the wind in your area is constantly changing in a circular way, or you do not intend to service or inspect your mill for weeks at a time, the simplest solution to the yawing problem is to fix several yards of cable to your generator and leave it loose. As the mill moves round on its mounting the wire may wind itself round the pole or stand but as long as you make sure it cannot foul the sails it will probably unwind itself just as readily, and if it shows signs of winding itself round the pole

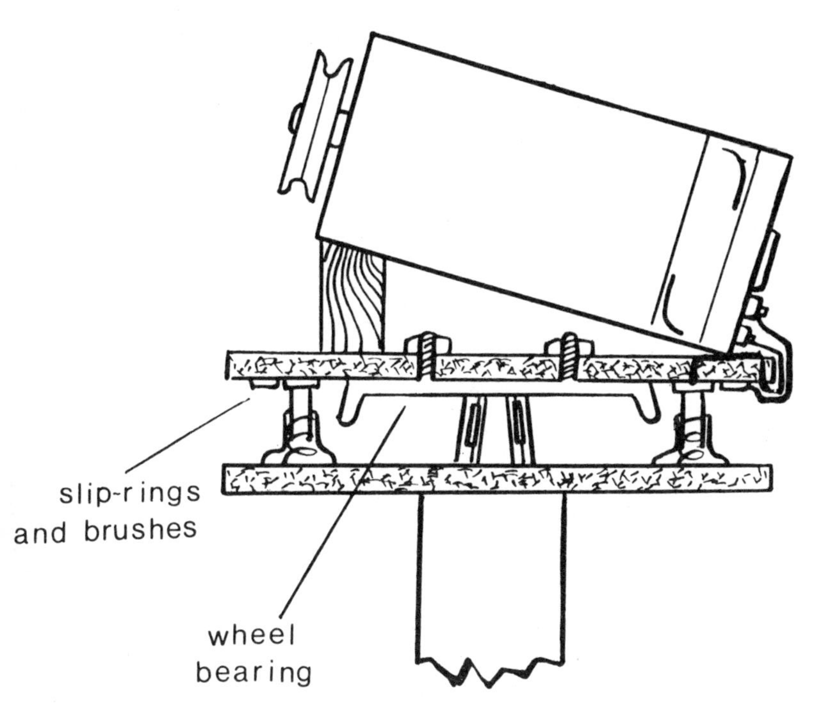

16 Mounting a generator on a rotating head with slip rings and brushes

too tightly you can move the whole mill around by hand a few times to undo it.

The best method is to fit slip rings, sliding electrical contacts. A simple way to fit slip rings to the mill described above is to screw two concentric rings of copper sheet to the moving disc that holds the mill, underneath the disc, and take the wires from the generator down through holes to be soldered to these rings.

Then, on the fixed disc, fit two carbon brushes as used for vacuum cleaner motors (you can buy these as spares very cheaply), and set them upright in a short length of plastic tube fixed to the stationary disc, so placed that they rub against each of the copper rings. You can then take off your power from these brushes no matter how often the mill spins round to face the wind.

Last, and perhaps most important, you will have to fit a tail to the upper disc to make sure that the mill does spin to face the wind. Get a 6ft length of steel conduit and bend it so that the final foot is at about 22° to the rest. Fix this end to the moving (upper) disc with conduit straps so that the other length points slightly upwards and directly back from the line of the mill axis. Now cut a slot in the far end of the conduit and fit an

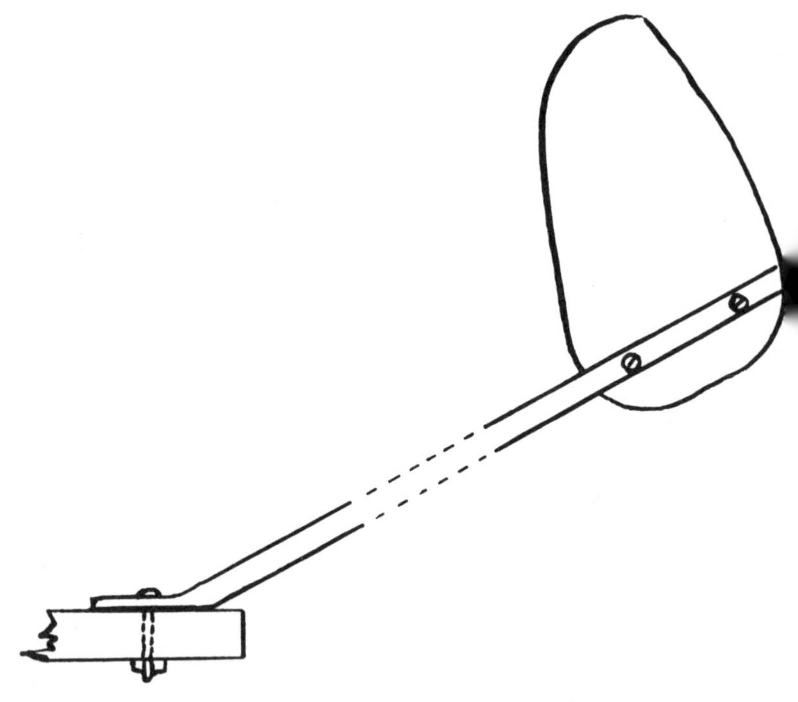

17 Tail for windmill. The tail can be made of aluminium sheet set into a slot in the tail arm, and bolted on

aluminium sail or rudder (*see Fig 17*) into the slot, fixing it with two bolts through the conduit.

Multi-vane fans

These fans or windwheels are used around the countryside as wind pumps. They consist of a ring of about ten to twenty fan blades set at about 25–30° to the plane of the wheel.

The multi-blade fan is even slower-moving than the slat-type mill sail, but it does start moving at very low wind speeds (the cut-in speed) and is therefore useful if the site is rather too sheltered. Like the slat-type sail it can be geared up to give a satisfactory performance.

One way to make a small multi-vane mill is to take an ordinary bicycle wheel and fit small vanes of aluminium at intervals round it, using the spokes as a guide. If you cut small

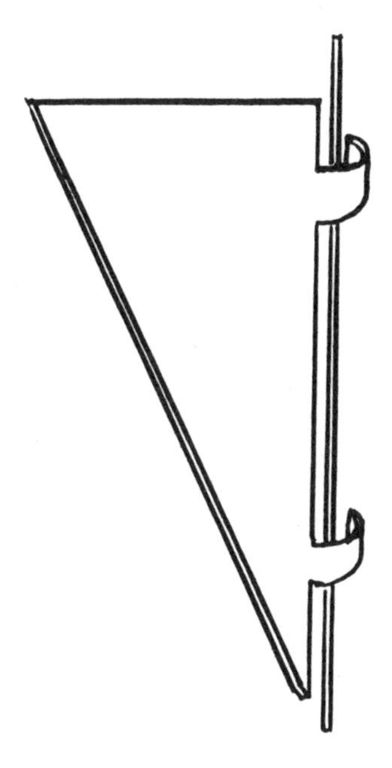

18 Simple aluminium vane to fit to bicycle wheel. The small lugs can be bent round the spokes and fixed with pliers

lugs on the vanes (*see Fig 18*), these can be turned round the spokes to hold the vanes, and wire spacers can be used to keep the aluminium at the right angle to the wind—about 30° to the plane of the wheel is suitable. A 20in bicycle wheel should be capable of driving a dynamo up to about 60W: a 25mph wind will give 50W. Alternatively you can use a wheel with a hub dynamo, which saves any constructional work. This will not, however, give as much power as the wheel can provide at best.

A commercial version of the multi-blade fan (*See Appendix 4*) is used to provide power for harbour beacons and other kinds of navigational light. Its fan blades are made of plastic and fixed to a rotating hub by a pressure ring. The device will charge batteries for caravans and boats.

Propellers

Propellers have the great advantage for electricity generation that they revolve much faster than slat-type or multi-blade designs, and therefore need less gearing to drive a generator at the right speed. Often the tip speed of a propeller blade is as much as six times the wind speed, so that in a 30mph wind a 6ft propeller will turn at about 840rpm, enough to drive a suitable generator without gearing up. For this reason propellers have been used extensively for the larger wind-powered electrical installations. Sometimes the speed at which propellers run can be a positive disadvantage for very large units, because the tips of the blades can approach the speed of sound when the propeller is turning in a high wind. This imposes enormous strains on the structure, and may be the reason for some of the breakdowns of large wind generators. Even with small propellers the blades must be strong and well-balanced.

The simplest way to make a propeller is to take an ordinary car fan, drill two holes in each of the fan blades, and bolt pieces of plywood to them. It may be necessary to bend the fan blades slightly more at an angle: a good weathering angle is 35°. You can do this easily by fixing the blade in a vice and gripping the centre of the fan between two pieces of wood, which give more leverage.

The plywood blades should be about 4in wide. I do not advise making them more than 2ft long, as the strains on even a 4ft propeller are quite high, and a strong wind may send a section of plywood flying through your window—or your neighbour's, which is worse. Balance the propeller by placing the centre of it on a broom handle or similar piece of wood; if it shows signs of imbalance either cut a small piece of plywood off the heavy side or put a nut and bolt through the light side. You can find the right place to do this by resting the nut and bolt on the surface of the plywood, moving them to and fro until the propeller balances, and drilling a hole at this point to fix them permanently.

If you want a larger propeller, it should be made of stronger material than the simple plywood model just described, and should be nearer the ideal shape. The pitch of a propeller blade should preferably increase from a low angle—almost flat to the wind—near the centre to about 45° near the tip. If you imagine looking along the blade from the centre of the propeller it will gradually turn backwards like a screw with a very large pitch; this shaping makes sure that the force of the wind is evenly applied all the way along the blade, although the tip is moving much faster than the centre.

The best way to make such a propeller, both for strength and to get the complex shape right, is to use glass-reinforced plastic (fibreglass). A moulded propeller would obviously be the ideal (and GRP propellers are available ready-made from one or two suppliers), but making a mould is a long and complex job.

The alternative is to make a steel skeleton and cover it with GRP.

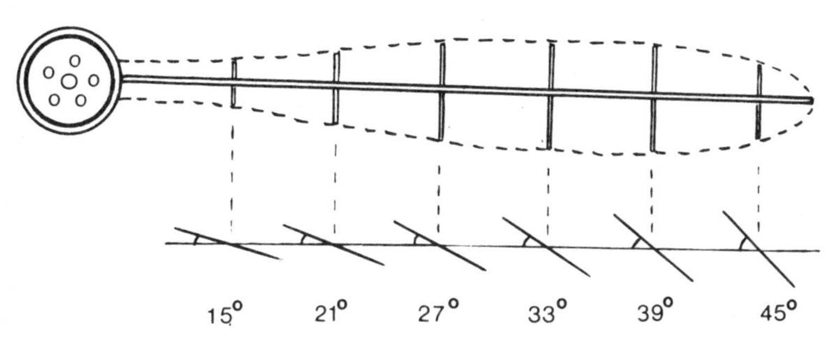

19 Skeleton for propeller blade. The lower drawing shows the correct angles for the cross-pieces so as to achieve the 'screw' shape

Figure 19 gives the dimensions. If you can get hold of it, concrete-reinforcing steel rod would be ideal for the 'spine' and fencing wire for the cross pieces. Solder the cross pieces where indicated, making sure that the angles are correct (if necessary, tie them into place with thin wire before soldering). Welding will make a quicker and stronger job of the skeleton, but the strength of the final propeller depends mostly on the GRP, not the steel, so do not worry if that is not possible.

When the skeleton is complete, clean it thoroughly to remove soldering flux, scale, rust and so on, and paint it over with rustproofing liquid (twenty-five per cent phosphoric acid). Leave it to dry.

Now run glass tape (woven glass rovings) up and down the skeleton until the shape has been filled in. You can make sure that the tape follows the correct angles by passing it in and out of the cross-bars. Mix up polyester resin and catalyst as in the manufacturer's instructions, and paint the glass tape with resin until it is completely impregnated. Leave it until the resin just begins to gel, then build up the shape with small pieces of glassfibre mat impregnated with the resin, brushing each one down well to avoid air bubbles or gaps between the

20 Approximate shape of fibre glass propeller when complete

layers. Figure 20 shows the shape to aim at: you may not be able to do this as smoothly as if you had a moulded propeller, but with care you can make a good approximation.

You will probably have to leave the work between layers to allow it to set up to the gel stage: remember that the resin will inevitably set once the catalyst has been added, so do not make up too much at a time. You can clean the brush and your hands, if they have become covered with resin, with acetone or a proprietary cleanser sold by the manufacturers. Acetone is highly flammable, and the resin is somewhat flammable, so take care.

Once the propeller has completely set, rub it down with emery paper, or use a sanding disc on a power drill (alumina discs are best) to give a smooth surface. (Good ventilation is necessary when you are rubbing down.) Drill out the fixing holes in the centre boss to take bolts.

You can, of course, make a three- or four-bladed propeller by a simple extension of this method. You will find that these cut in at lower speeds than the two-bladed type, so they offer some advantage if your site is not very windy.

The propeller needs mounting on a shaft in the same way as the slat-type mill already described. You can either make it drive the generator directly or fix it to a pulley wheel (a car fan pulley is a good size) and drive the generator with a belt. You will not need such a high step-up ratio as for the slatted sails—5:1 should be ample for most generators.

The whole mounted propeller and generator has to be able to yaw into the wind, and the instructions already given for mounting should be followed. Again, you can use loose wires and 'wind' the mill back when they get twisted, or go to the trouble of making slip rings to carry the power from the generator.

General

All horizontal-shaft mills have a critical wind-speed or 'cut-in', below which they do not start operating. Multi-blade fans have the lowest, and will frequently start moving when the wind speed is as low as 6mph. Slat-type sails can also use fairly low wind-speeds. Of course, as can be seen from Table 5, not much power comes from a windmill at such low speeds, but there may be enough to trickle-charge a battery. Propellers need higher speeds of wind to start moving, and many will not operate until the wind is around 15mph or more.

Gales can do a lot of damage to the mill and, if it is overrun, to the generator. Some complex mills have devices for 'feathering', altering the pitch of the blades so as to offer less obstruction to a really strong wind, but obviously such devices are complicated to construct. The alternative is to turn the whole mill out of the wind so that the sails or propeller no longer rotate. Again there are automatic devices for doing this, but you can do almost as well with a length of nylon cord tied to the tail of the mill and a stout tent peg to secure it to the ground. If you expect a gale, tie the mill down so that the propeller or sails are in line with the wind, not facing it.

As an extra precaution, you could also fit an extra pulley (a car fan will provide a suitable size), over which you can pass a length of hemp rope (nylon would melt in the heat from friction) to slow down or stop the mill in high winds. This is also useful if you want to stop the mill for servicing or lubrication —a mill sail carries a lot of force behind it and could hit you painfully hard if you tried to stop it with your hands.

Vertical-shaft mills

Most people will have seen advertising signs which have a circular sheet of metal which rotates when the wind blows. These are elementary vertical-shaft windmills.

Mills of this kind have one great advantage over the horizontal-shaft type—they do not have to yaw into the wind, because they are symmetrical and can receive it from any direction. On the other hand, they have a correspondingly serious disadvantage; looking at Figure 21a, if you imagine that the cross-shaped mill is revolving because face A is being blown forward by the wind, it is clear that face C is having to move *against* the wind, so that the rotation is severely damped.

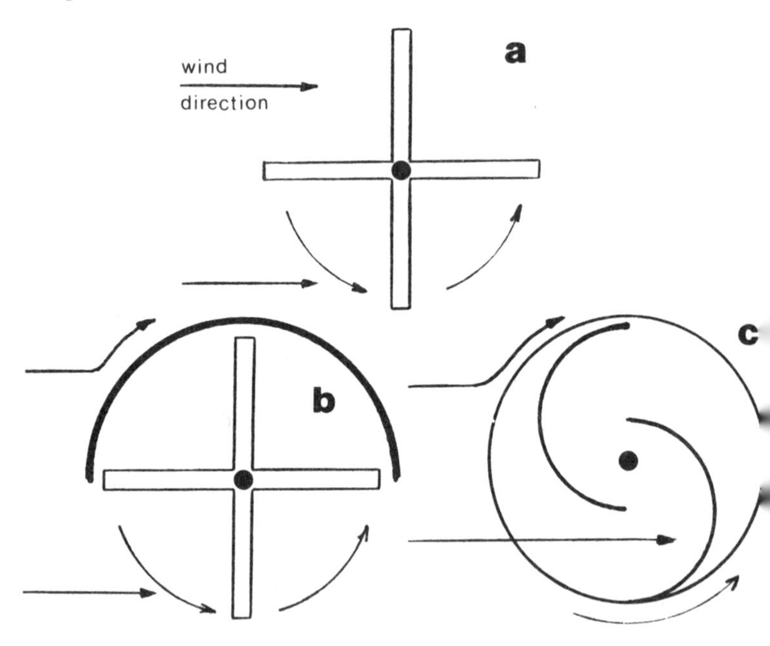

21 (a) Simple vertical-shaft mill, (b) Vertical-shaft mill with baffle to divert wind, (c) Savonius rotor

Various ideas have been put forward to overcome this. Figure 21b shows a kind of shelter built round half the mill to protect the faces that are moving into the wind from its pressure. Unfortunately such a shelter would have to move round as the wind veers, so that we are back to yawing again, and there is no real advantage for the vertical-shaft design. Robert Beatson, in 1798, described an ingenious vertical-shaft mill in which the vanes were made like louvred doors: on the vane being driven by the wind these were closed and therefore took the force of the wind, but as the face came round against the wind they opened and let the air pass through. This design was superseded by one put forward by the Finnish engineer Savonius (*see Fig 21c*). In this mill the two half cylinders, mounted off-centre between two flat discs that carry the bearings, act as the vanes. As each concave side comes round the wind blows into it and keeps the mill revolving. Obviously the convex faces have to come round against the wind, but because of their shape they are actually sucked back into the wind by an aerodynamic force known as the Magnus effect. This is similar to the lift which acts on an aeroplane wing at the curved surface of the upper wing passes through the air. Because of the Magnus effect, a Savonius rotor uses the power of the wind far more efficiently than other types of vertical-shaft mill.

You can make a Savonius rotor very easily by cutting an old 50-gallon oil drum in half down the middle, so as to produce two half-cylinders. Now cut two circles of ½in (12mm) plywood, 3ft in diameter, and bolt the half-cylinders to them (*Fig 21c*). Use large washers to spread the load on the bolts, as the metal at the ends of the drums is not very thick and could tear apart if ordinary bolt-heads are the only fixing. These discs have to be fitted with bearings. The lower bearing can be made from a wheel bearing, using the wheel studs to bolt the plywood disc into place. The upper bearing does not take much load, so a simple piece of shafting and a ball-bearing should suffice. Fix a pulley wheel to the upper shaft to take a belt drive.

The framework for the rotor should be made in 4in (100mm) square timber: this may seem heavy, but the forces on the rotor can be great in a high wind. The generator can be bolted to this framework at a safe distance from the rotor and driven by a belt from the pulley, or preferably a system of two belts to give a step-up ratio of about 20:1. The Savonius rotor is easy to build and maintain, but it must be understood that it does not run very fast even in a high wind, and it is not really as efficient as a horizontal-shaft mill, so that it cannot give the power that its size might suggest. However, for the home constructor, cheapness of materials and ease of construction are often primary considerations, and the overall efficiency of the system is secondary, as long as it works fairly well and does not need a staff of trained engineers to keep it running. A number of groups have designed models similar to the system described here, using oil-drums and similar simple materials, to make Savonius rotors for underdeveloped countries. They can be used for generating electricity, for pumping water and for other agricultural work, for which their slow speed is no disadvantage.

Another type of vertical-shaft mill has been designed by the National Research Council of Canada specifically for electricity generation, giving a faster rate of rotation than the Savonius rotor (*see Fig 22*). The rotor carries three narrow bands with aerofoil section, set at an angle to the wind. So far the research

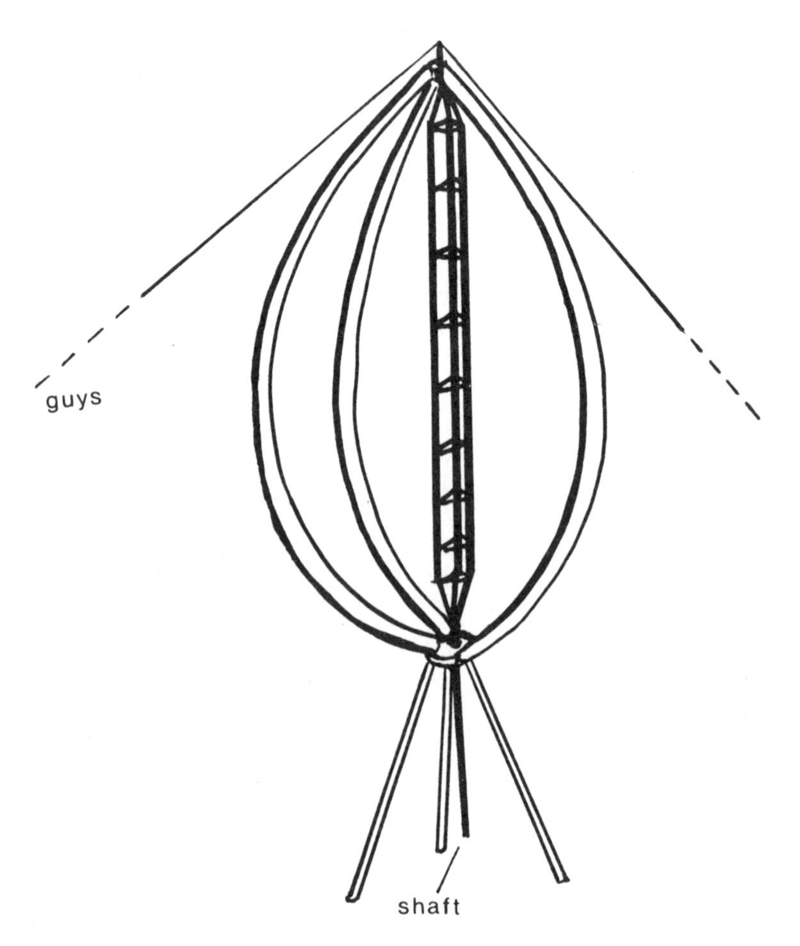

22 National Research Council of Canada vertical-shaft mill design

on this type of rotor has not been completed, but readers could experiment with variations on the design for small rotors.

Mountings

All windmills should be mounted as high as possible to get the best of the wind without interference from buildings or trees—which not only diminish the wind force, but cause turbulence that can make a mill run very irregularly. Also, high-speed propellers must be mounted high enough not to slice the heads off passers-by! Large professionally made mills are usually mounted on pylons made of cross-braced steel, and if you contemplate making a mill larger than about 6ft diameter to drive a large generator you will also have to think in these terms.

Smaller mills can be mounted on tubular steel poles or scaffolds: you can make up a suitable structure in scaffold tubing (steel or aluminium) or galvanized steel barrel, held by welding or with scaffolding ties, which can usually be bought at a builder's yard. The base of the tower should be set in concrete, if it is isolated from buildings, and it must be braced with four wire guy-ropes anchored into the ground and tightened with turnbuckles. Keep these greased and check them for tightness from time to time, as vibration can make them work loose. When fixing the guys, make sure that they cannot foul the mill sails or propeller, and use them to keep any trailing cables out of the way of the moving parts—this is especially important if you are not using slip rings to transmit the power but relying on a loose length of cable.

Instead of tubular scaffolding, slotted-angle framing (Dexion) can be used to make a narrow tower. This would be more suitable for a Savonius rotor which needs a rather square platform for mounting. Again you will need to brace the structure with guy-ropes.

If you are mounting your mill near a building, use it to support the tower or pylon. You can get boss-heads for bolting to a wall, which then hold the ends of tubular steel scaffolding bars, or you can bolt slotted-angle sections to the wall with Rawlbolts. Try to spread the load of the fixings as widely as possible so as to minimize the chance of pulling bricks out of the wall or cracking concrete: windmills vibrate a good deal at certain speeds.

If you have a flat portion of roof, not too overshadowed by chimneys or other roof sections, you can sometimes fit a mill there on quite a small mounting structure, bolting through to the rafters of the roof below. (Do not depend on simple wood-screws through to the timber cladding alone, unless the mill is quite small, but locate a rafter and use coach bolts right through.) Some people recommend lashing a mill to the chimney with wire and corner plates, as TV aerials are sometimes fixed. But in high winds even an aerial may cause chimney brickwork to give way; and a windmill has to be able to take far greater strains.

If you have a tall tree which is due to come down, do not chop it from the base, but cut off the branches down to the top of the trunk and leave this flat as a platform for a windmill. The roots of the tree will keep it completely stable. Remember however that the wood will shrink as it dries out, so fix your windmill with expanding bolts (Rawlbolts) in drilled holes, so that you can tighten them up from time to time.

Weatherproofing

The effects of weather on windmills are exaggerated because of the speed of movement. The tip of a propeller can easily be travelling at 100mph in a strong wind, and if raindrops hit it at this speed they hit hard. So paint every wooden or metal part very thoroughly with a tough, waterproof paint or varnish—polyurethane paint or marine varnish. Use grease on all the moving parts, such as bearings, and make sure that you renew it frequently.

The actual generators are obviously susceptible to weather, and should be inspected from time to time for dirt or rust. Very often they can be protected by a wooden box or even a plastic bowl bolted over them, with just room for the belt drive to pass through.

7 Power from water

While hydroelectric power makes a very important contribution to many national systems, only a few people have access to a water system that could be used to generate electricity. There are, of course, small water turbines that run from the tap supply and could be used to generate electricity from a small dynamo, but the use of purified water in such a wasteful way is too anti-social to consider.

If you do have a stream, check your legal position before you start any work on hydroelectricity, and that you would not interfere with the use of water by people further downstream. If you are content to have a fairly small generator, the water does not have to be a raging torrent or a 50ft waterfall.

The older type of mill-wheel with its large paddles fixed to a wooden wheel was ideal for turning millstones, but it ran very slowly and needed a mill-dam to hold up the water—which would certainly not be permitted now. For running an electric generator the more modern turbines are necessary: these work with a smaller amount of water than the large wheels, and therefore interfere less with the flow of the stream.

For large installations the Pelton wheel is popular: its efficiency in converting the energy of the stream into work can be more than ninety per cent. The Turgo Impulse Turbine, a development of the Pelton wheel, is as efficient, and develops large amounts of power with a relatively small turbine wheel. However, for most small installations the problem is that Pelton wheels really need a head of water of at least 100ft (30m).

The Francis turbine is better where the head of water is small: it has a runner (*see Fig 23*) which is normally set with a

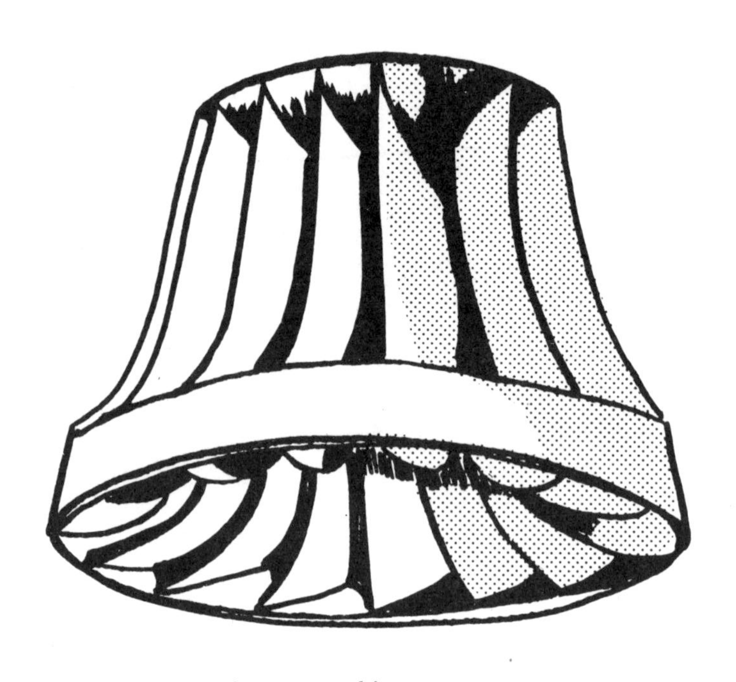

23 Runner for Francis water turbine

vertical shaft, the water supply running downwards through it. As the water passes through, the impellers on the runner turn it at 50–250rpm according to the head of water and the size of the turbine. Francis turbines can run reasonably well on a head of water of as little as 5ft (1.5m), while some turbines used in large hydroelectric schemes can cope with a head of water of 1,640ft (500m) and generate 83,000HP.

The size of the runner of a Francis turbine depends on the amount of water available, and also whether the turbine is to be run at slow or fast speeds. Obviously if x gallons of water per minute are to pass through the turbine, they can be accommodated by a narrow but fast-moving runner or by a wider, slower one. High speeds are desirable for running an electric generator, but on the other hand efficiency goes down if the rotor is too fast: for a head of 20ft (6m) for example, 125rpm is about the best speed. In general, before allowing for the various losses due to friction and irregular flow of water, the output of a Francis turbine is proportional to $H^{\frac{3}{2}}$ where H is the head of water. Commercial Francis turbines are available, some already fitted with suitable generators, others needing adaptation for electricity production.

The hydraulic ram

As the head of water is so important for running a water turbine, any device that can raise water to an artificially high level will benefit the turbine system and extract more energy. A turbine that gives 1HP at a 10ft head will give over 5HP with a 30ft head of water.

Obviously pumps could be used to lift water into a high storage tank to get extra pressure, but this would be a self-defeating system because the power needed for the pumps would be more than any power that could be extracted from the water. However, if the water is to be raised from a flowing stream, there is a mechanism which makes use of the energy in the whole mass of moving water to lift a small proportion of it, without any external supply of power. This is the hydraulic ram.

Essentially a hydraulic ram consists of a pressure vessel into which water runs from the main flowing stream. A loaded valve lets most of the water out again, but not at the same rate, so that a head of water builds up in the pressure vessel. When this head of pressure is sufficient to shut the valve, there is a check in the flow and the pressure is transferred with great force to an air vessel. The air in this compresses, and the force so set up is sufficient to drive a small amount of the water up a pipe to a much greater height than the main head. As it goes it gradually releases the pressure, the pressure-loaded waste valve opens again, and the process is repeated. The action consists of a fairly steady series of compressions suddenly interrupted at intervals by spurts of water up to the higher level. Obviously the higher the level to which the water has to be raised, the more time it will take to build up sufficient pressure, so the spurts will be shorter and the build-up time longer.

Figure 24 shows the process diagrammatically. If the head of the main supply is $h$ft (or h metres—it does not matter how the figures are expressed as long as the same units are used for all the heights), and the flow coming into the ram is F gallons per minute, then the amount of water (in gallons per minute) pumped up to height H above the ram will be $\frac{2 \times F \times h}{3 \times H}$

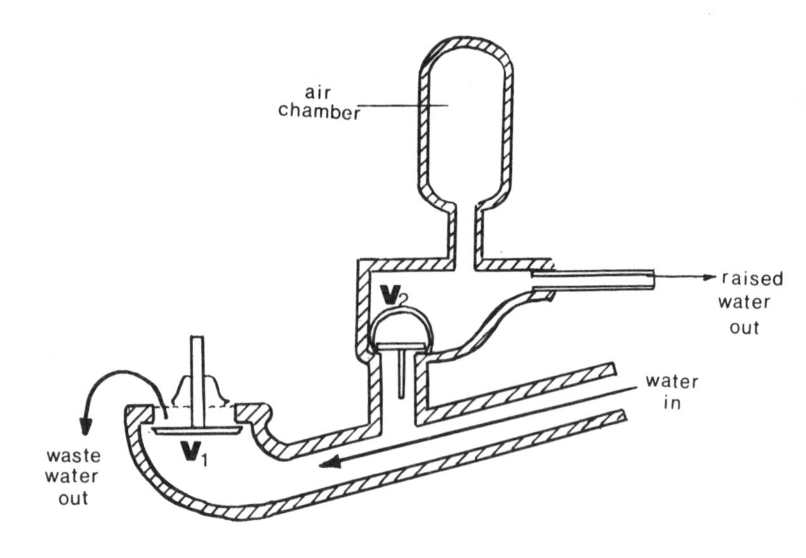

24 Hydraulic ram. V_1 is the waste valve and V_2 the valve to the air chamber and lift pipe

approximately. This formula will work as long as H is not more than about 5 times *h*.

For example, suppose that the working head of a stream is 5ft, and it is possible to get a flow of 4mph down a six-inch main to supply the hydraulic ram. Then $h = 5$ and $F = 430$gal/min. If we want to raise some of this water to 30ft above the ram, then the rate of delivery will be $\frac{2 \times 430 \times 5}{3 \times 30} = 48$gal/min. The remaining 382gal/min will return to the stream having given up some of its energy to lift the rest.

The water so raised can be used to run a small turbine (a quite small one in this case), and the water from the turbine returned to the stream. In this way no essential interference in the water-flow is caused by the turbine action.

The hydraulic ram needs no power, and as, apart from the pressure valves, it has no moving parts, it needs very little maintenance either.

8 Power from the sun

When craft such as Skylab 3 are wandering through the almost empty spaces between the planets, they are cut off from most conventional types of energy, yet their usefulness depends on the ability to send back TV and other messages. Batteries add weight, and cannot go on producing power for long periods. The solution lies in those curious flaps, like wings, that the spacecraft put out as soon as they have reached their orbit. The flaps are in fact arrays of solar cells, which take in the sunlight and convert it to electricity. As the spacecraft are exposed to the full heat and light of the sun, not diffused or shadowed by air or clouds, a lot of ènergy is available if it can be used efficiently. In fact some satellites, such as the communications satellite Intelsat 4, may be using as much as 900W generated directly from sunlight.

It has been known for many years that some materials produce an electric current when exposed to light. For example, cadmium sulphide cells have been used for years in lightmeters for photography, but even in a strong light such cells only produce milliwatts of power, enough to feed into an electrical measuring system to assess the light level but no use as a primary source of power. The arrays of solar cells used in spacecraft have been costing vast sums, but the efforts being made to mass-produce them are now bringing their price into the realms of possibility for more ordinary jobs.

There are two main types of cell that are important at present, the cadmium sulphide cell and the silicon cell.

The cadmium sulphide cell, as has been said, has been around for some time as a light-detecting device; many are used in lightmeters and in electronic devices such as intruder alarms and counting systems on production lines. If you shine a beam of light, even the invisible infra-red light, on a cadmium sulphide cell it will produce a small current. By amplifying this current you can make it connect to an alarm, a counter, or something similar, and as soon as anything or anybody interrupts the beam of light, the alarm or counter operates. So far the cells have had rather low efficiencies, but large cadmium sulphide cells on sheets of plastic, big enough to take in a lot of sunlight, and therefore able to produce usable amounts of electric power despite the small yield per square metre, have been developed recently. Cadmium sulphide is a cheap material and once mass production is begun these solar panels would be a commercial proposition for domestic use.

Silicon cells are already available in the form of arrays of small cells mounted on panels. Although expensive compared with other methods of producing electricity, their price is coming down.

The heart of a silicon cell is a very thin slice of pure crystalline silicon treated with small amounts of phosphorus and boron. The manufacture is similar to that of a transistor for ordinary electronics work, but as the solar cell has to absorb as much light as possible, the tendency is to make larger and larger slices of silicon, as technology permits, while of course in the conventional electronics field the aim is to make everything smaller and smaller.

The amounts of phosphorus and boron and the purity and crystalline form of the silicon are critical, so it is difficult and expensive to make large cells. Because the current and voltage from each individual cell are very low, they have to be put together in a series-parallel array (*see p 88*) to get a suitable amount of power for battery charging.

For instance, the Lucas Solar Battery Charger consists of a set of thirty small circular cells arranged in a panel 18in × 14in (450mm × 345mm). This panel will generate a voltage a little over 12V, enough to maintain the charge of a car battery, at an average rate of 0.6A during daylight, giving an average rate of charging of about 25Ah per week under temperate conditions. (For comparison, an ordinary car battery usually takes about 50Ah for full charge, so a boat or caravan used every two or three weeks could always have a fully charged battery ready if one of these panels was connected.)

Solar cells need no other fuel and no servicing, have no moving parts to wear out or break down, and only need to be connected to a battery to produce usable electricity. For smaller loads the panels or cells can be used direct: simple devices for running transistor radios from one or two solar cells, for example, are popular in Australia and the USA. However, as the power disappears at night some form of storage is desirable.

Panels such as the Lucas model are designed to stand up to stringent conditions, such as motor fumes, salt water and extremes of heat and cold, and only need occasional cleaning to keep them in running order. They are therefore suitable for boats, light-buoys, beacons, and other isolated situations.

Available power from the sun

It is not always realised just how much solar energy is available, even in temperate climates. People accept that solar power can make an effective contribution to the energy supply in New Mexico, but tend to assume that in areas with clouded

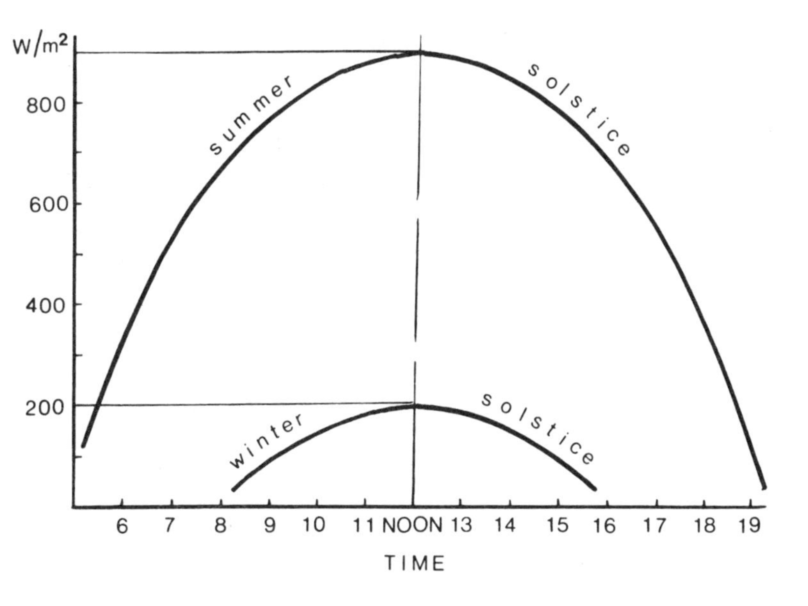

25 Average sunlight energy reaching Britain during a day in midwinter and midsummer (watts/metre²)

skies it is insignificant. Yet look at the average radiation received from the sun over the British Isles, in watts per square metre, on 21 June the summer solstice, and 21 December the winter solstice (*Fig 25*). Around noon the sun is strongest and radiates most energy, while this radiation falls off nearer dawn or dusk. It will be seen that at the height of summer, around midday, we could theoretically collect as much as 900W per square metre, and even in winter 200W/m² could be obtained. If we add together the variable energy supply over the whole day and average this supply over the year, summer and winter, we find that each square metre receives an average of 2½kWh ('units') per day. (This figure allows for the losses due to low cloud, smoke and other pollution, and similar things which tend to obscure the sun.) Of course the figure for summer will be much higher than the average, but even in winter the supply rarely falls below ½kWh/m² per day.

Assuming an average household consumption of about 50kWh per day, this could be collected on a panel of 20m² (say 20ft × 11ft), if an efficient collecting system were available. In practice no systems are 100 per cent efficient, but simple flat-plate collectors have achieved about 50 per cent of the theoretical figure. (These are hollow blackened plates filled with water which act rather like a central-heating radiator in reverse, taking in heat from the sun to produce hot water for the household). Assuming this efficiency, a panel 40ft × 11ft could provide all the power for a household, just from the sun.

Where the power is to be used as heat—for central heating, greenhouses, hot water for baths and washing, and so on—the solar power would be used to heat water directly and pump it where it was needed. The existing solar electricity generators are not really as efficient as simple solar water heaters—if you put up that 40ft × 11ft assembly of solar cells it would cost you, at present prices, about £50,000 (or $90,000), and you would get an average of 12kWh per day, only about a quarter of the necessary supply. However, research in this field goes on so fast that this position could soon change.

Setting up a solar panel

If you decide to buy one of the commercial solar panels, for an emergency supply of electricity, or to provide power for a boat or caravan, you can much improve its performance by care in the mounting. Clearly the panel must face the sun for as long as possible each day, but unless you are going to build a complicated mechanism such as astronomers use to keep their telescopes pointing at one part of the sky, you will have to decide on an angle that gives the best average performance.

First, the panel should face south (or north if you live in the southern hemisphere!). The actual parth of the sun varies during the year (it is only at the equinoxes that it goes precisely from east to west) but on average it is at its strongest when shining from the south. Calculation of the angle that the panel should present to the ground is complicated, but a fairly simple result emerges: solar energy is best collected by pointing the panel surface about 10° higher than the lowest noon solar point in midwinter. (This is because the coldest time of the year is actually after the winter solstice on 21 December: otherwise the panel would be placed pointing directly at the lowest noon point). If L is the latitude of your site, in degrees, the angle of the panel to the level ground should thus be $(L + 23\frac{1}{2} - 10)°$ or about $65\frac{1}{2}°$ for places on latitude 52°N, such as Bedford or Worcester (UK).

This angle gives the best result for making use of the sun in winter, when most people would want to extract as much power as possible. However, a great many of the uses for solar-panel electricity only arise in summer, for holiday caravans, boats and so on. Then the sun will be higher in the sky, on average, and you should set your panel back further. For summer months the best efficiency will be obtained if the angle of the panel to the horizontal is $(L - 23\frac{1}{2})°$ where L is the latitude as before. (The calculation has been much simplified, and it will be obvious to the reader that these formulae lead to strange results if used near the equator. The answer is, quite simply, that if you get lower then latitude $23\frac{1}{2}°$ your panel should be horizontal.)

Another improvement to the simple panel is to increase the amount of sunlight that falls on it by placing suitable reflectors around it. Figure 26 shows a simple arrangement made with sheets of hardboard covered with aluminium cooking foil as reflectors: the aluminium is painted with clear polyurethane varnish to protect it from the weather. A simple calculation

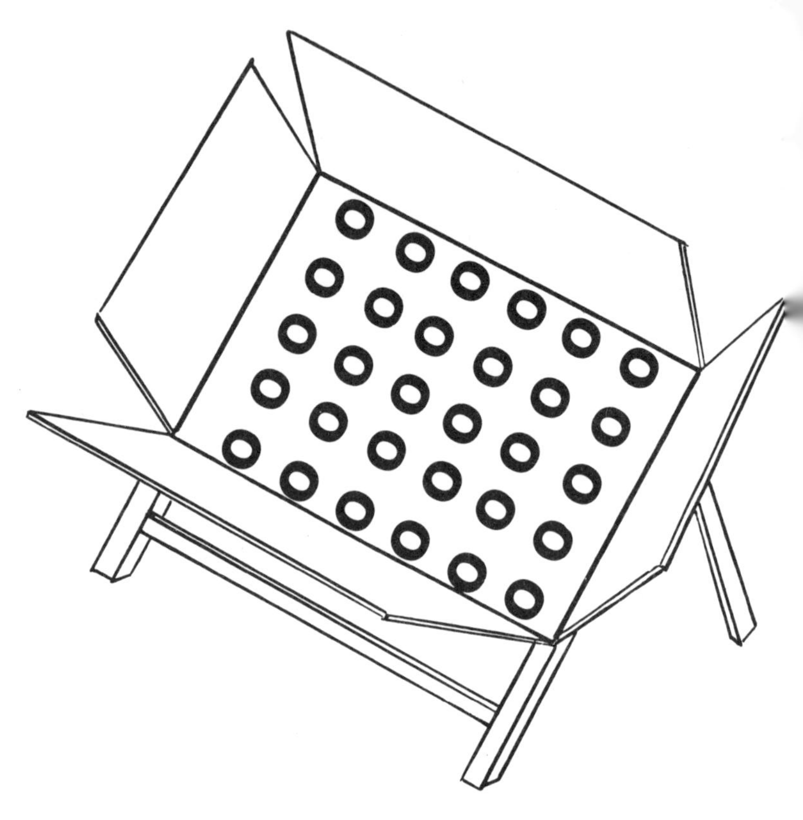

26 Improving the performance of a solar panel by surrounding it with reflecting surfaces at 60°

shows that the best angle of the reflectors to the panel is 60°: the reflectors should preferably be the same size and shape as the panel, as this gives the maximum reflected light.

9 Thermoelectric power

While most of this book deals with the production of reasonable amounts of power for home use, say 100–2,000W, using conventional means adapted for the home constructor with limited resources, it seems necessary to mention some of the more unorthodox and experimental methods of producing electricity, even if at present they can only be used to make tiny amounts of power. Technology changes so fast that ideas only on the drawing board as I write may be practical propositions by the time the book is read.

Thermoelectricity

Most electricity comes from heat in some way: we burn coal or oil in boilers and use the steam to drive generators, we extract the heat from radioactive material and use it in the same way. Even wind power is largely derived from the air movements caused by the heat of the sun, and the greatest source of natural electricity, the lightning flash, is the result of clouds moving past one another in these air currents and thus also moving enormous charges of static electricity.

The trouble with most of the methods is that they are wasteful of heat. Even our best power stations burning coal or oil waste nearly seventy per cent of the fuel value: the heat disappears into the air or cooling water.

There are methods of converting the heat directly into electricity, without going through the wasteful stage of using it to drive an inefficient engine which in its turn drives an inefficient generator. One of these methods is the use of the Seebeck effect.

The physicist Seebeck discovered in 1821 that if he made up a circuit using wires of different metals, say iron wire in half the circuit and copper wire in the rest, odd things happened. For example, if he heated one of the joining places and left the other cold, current would flow from the copper to the iron at the hot junction, and round the circuit. The voltage produced was very small (only 0.0013V when the hot junction was at 100°C and the cold one at 0°C, but it went up as the difference in temperature was increased, until the upper temperature reached 275°C. At higher temperatures than this the voltage diminished again. Higher voltages could be got out of the system by connecting together a number of pairs of wires in series (*see Fig 27*). If there were twelve pairs of junc-

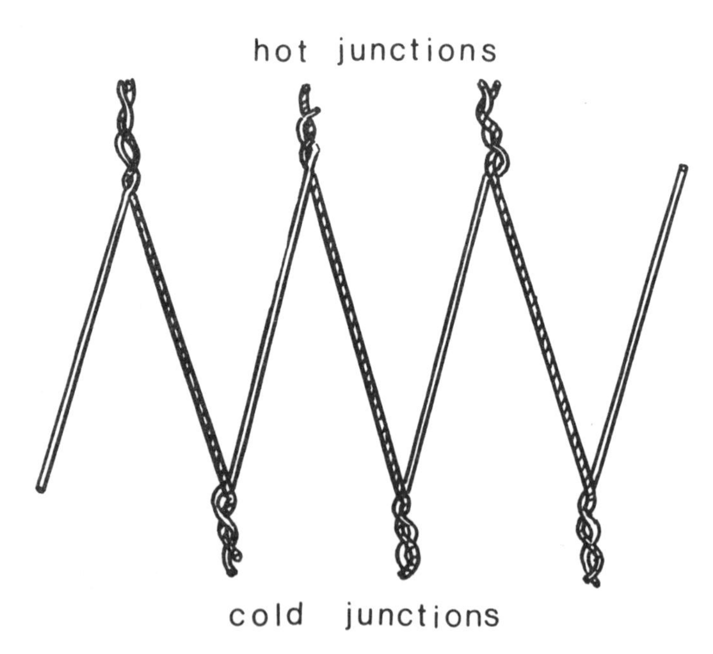

27 Principle of thermocouple. If one set of junctions is kept hot and the other set cold, a current will flow

tions, for example, they could get 0.0156V for a 100°C temperature difference.

Such a system forms the basis for the thermocouple, a useful device for measuring high temperatures in places where ordinary thermometers would boil or be too easily broken. A set of pairs of wires is arranged so that one end, the cold end, can be kept in a freezing bath or some other cold place with a constant temperature. The other set of junctions is then brought into contact with the heat to be measured and a voltage flows between the wires and round the circuit. This can be measured and used to calculate the temperature.

You can make a thermocouple of this kind with iron and copper wire. Fix the wires together by twisting at the ends, and make sure each wire is insulated from those round it so that the current cannot leak away: plastic sleeving is obviously good for this, but makes the bundle very bulky if you have many wires. Thin pieces of polythene or other plastic can be woven between the wires as they are connected together, and this makes a more compact bundle.

Of course, one of the difficulties with the iron/copper combination is that it takes about 1,000 pairs of wires to give over 1volt unless you are working with quite high temperatures. Seebeck and others found that other pairs of metals could produce higher voltages per junction, and eventually drew up a series: if you make up junctions between any metal at the top of the column and any other metal below it, the current will flow from the 'upper' metal to the 'lower' metal at the hot junction, and the further apart the metals are in the column, the larger the voltage produced will be. For example, with bismuth and antimony wires and a 100°C temperature difference between the hot and cold junctions, the voltage produced would be about 0.014V, ten times as much as with the copper/iron system.

Table 7 Thermoelectric series

Silicon
Bismuth
Nickel
Cobalt
Palladium
Platinum
Uranium
Copper
Manganese
Titanium
Mercury
Lead
Tin
Chromium
Molybdenum
Rhodium
Iridium
Gold
Silver
Aluminium
Zinc
Tungsten
Cadmium
Iron
Arsenic
Antimony
Tellurium
Germanium

Some of these materials, although attractive in terms of the voltages that could theoretically be obtained, are not really usable in any simple couple. Silicon and germanium, for instance, should produce a high voltage if made into a couple, but in practice germanium is very expensive to make into wire, and silicon conducts rather badly, thus cutting down the current. Platinum, rhodium, iridium and palladium are all extremely costly, although these metals are in fact used in thermocouples for measuring very high temperatures as they can all tolerate hot conditions without melting or failing.

Some of the more exotic metallic couples may be made useful for electricity production by the discovery that some combinations of the materials can be made chemically, rather than by the laborious method of fixing thousands of wires together. For instance, bismuth and tellurium, from their positions near the top and the bottom of the Table, would be expected to give an effective thermoelectric voltage. In fact, crystals of the compound bismuth telluride, made from the two metals, give a remarkably high thermoelectric yield if heated at one point and cooled at another. So far the crystals are not readily available, and the most important industrial application has been to use the Peltier effect, the reverse of the Seebeck. If you pass an electric current round a circuit made of two dissimilar metals, as in the thermocouple, one junction gets hot and the other cold. Crystals of bismuth telluride have been used for Peltier cooling in refrigerators—a current passed through two of the crystals makes one hot and the other cold.

However, at least one commercial device produces electricity by the Seebeck effect—a small Russian generator intended to power radios and similar equipment in the very isolated regions of Siberia. This draws its heat power from the chimney of an ordinary oil lamp, and produces enough to run a transistor radio.

There has also been a suggestion from Sweden, though not yet a commercial proposition, that cars could be run more efficiently if the heat from the fuel was supplied to thermoelectric crystals such as bismuth telluride, and the electric power used to drive a motor.

Work on various thermoelectric combinations (as suggested by the various metals in Table 7) is going on in several laboratories round the world, so it should not be long before commercial thermoelectric generators are available. These could be used either to make better use of conventional fuels, by applying the heat directly to thermoelectric crystals, or even to catch the heat from the sun and focus it on the crystals with lenses, Fresnel lenses, or a system of mirrors.

10 Storage batteries

Most of the electricity sources we have considered are intermittent or irregular in some way. Even a professionally made
petrol or diesel generator must be stopped for refuelling
cleaning, lubrication and other servicing, as well as unscheduled
breakdowns, and if you depend on the sun or the wind for
power you will soon learn what Shakespeare meant by 'the
uncertain glory of an April day.'

Storage of electricity is in fact a serious problem even for
national supplies; at the moment power-stations have to be
big enough to supply the whole of the peak demand, and
consequently are grossly under-employed during off-peak
periods in mid-afternoon and early morning. If electricity
could be stored on a large scale we could manage with much
smaller power-stations run more efficiently. One idea is to
pump water uphill to large reservoirs during surplus periods
and then use this to drive hydroelectric power-stations at peak
periods, but this demands the right sort of geographical
situation (and also means duplicating the power plant).

Storing electricity in batteries has three main advantages
for small-scale usage.

1. The batteries keep working even when the primary
source is running slowly or even cut out: this will even out the
supply from an intermittent source such as a wind generator.

2. If enough batteries are available, they can be charged up
during periods of low demand, and used to run, for a limited
period, more equipment than the primary source could power
direct. For example, if you have a wind generator that can
supply an average of 500W, and store all the electricity, you

can use this to run equipment up to 2000W as long as you confine its use to six hours per day or less. In practice this means that you can have a good level of lighting every evening with a fairly low power source as long as it is running all day.

3. Most power sources give a variable voltage according to the conditions—the speed of the wind or the brightness of the sun, etc. On the other hand, most equipment needs a constant voltage, and may not work at low voltages or may be damaged by high voltages. Batteries can be charged from the variable source and will then give a constant voltage supply.

The disadvantages of batteries are their cost and the amount of space they take: this is why there is, up to now, no large-scale storage system for national electricity supply.

Types of storage battery

In storage batteries, as in dry batteries used in torches and transistor radios, the electricity is produced by a chemical reaction. The dry battery, however, stops working when the reaction is complete—in the commonest type of dry battery, the zinc casing dissolves in ammonium chloride to make zinc chloride. In an exhausted unsealed battery you can often see where the zinc has been eaten away.

In a storage battery the chemical reaction goes on, but it can be reversed if electricity is passed through the battery, so that the chemical reaction is also reversed and the system is ready to make electricity again. In the commonest type of storage battery, the lead/acid battery used for cars and many other purposes, lead and lead dioxide dissolve in sulphuric acid, to make lead sulphate, as the battery supplies current. When all the lead dioxide has been used up the battery stops working—it is 'flat'. However, if you pass electricity from a battery charger through it at this stage, the lead sulphate splits up into lead, lead dioxide, and sulphuric acid again, and the battery is ready to supply current once more. Other storage batteries are made with other metals, but in each case the chemical reaction is reversible, so that it can be used as a way of storing electricity until needed.

Lead/acid batteries

This is the ordinary car-battery type, and is the easiest storage battery to obtain. The system dates back to 1859, when Gaston Planté passed electricity through two lead strips in a jar of sulphuric acid, and found that not only did they change colour, one strip becoming almost black with spongy lead, and the other brown with a layer of lead dioxide, but he could get electricity back from the system when the incoming supply had

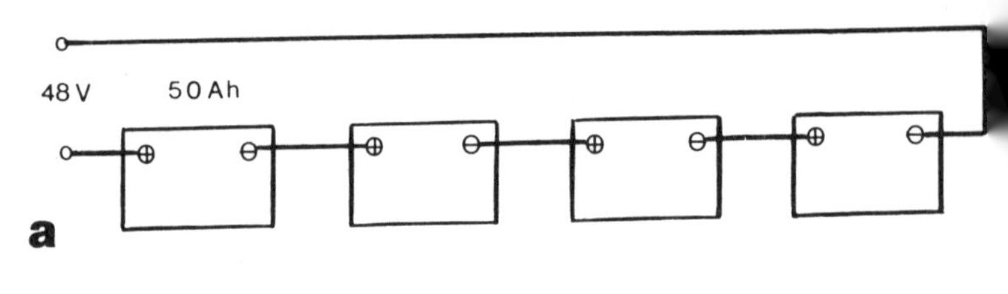

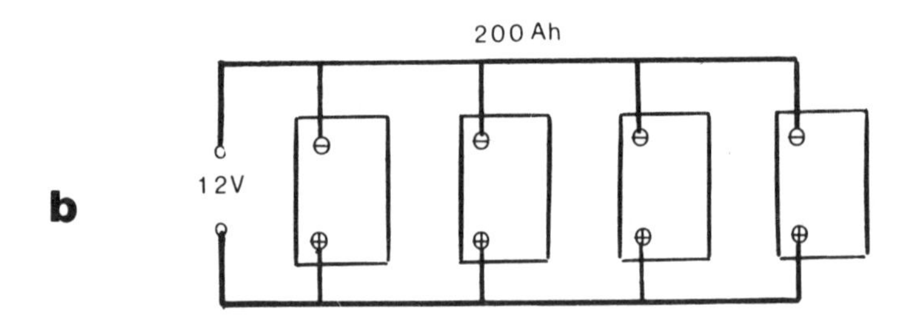

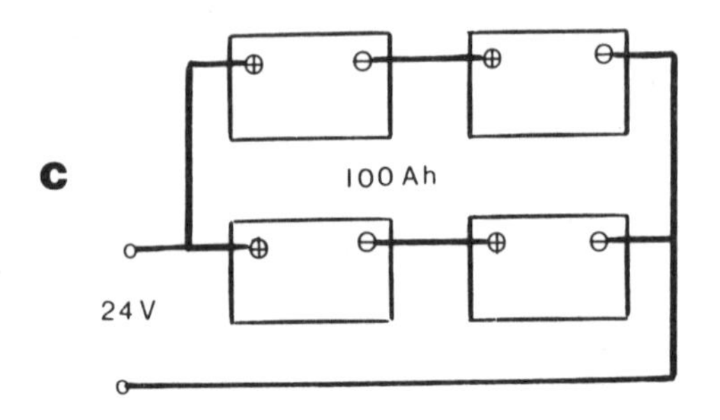

28 Batteries arranged in (a) series, (b) parallel, and (c) series-parallel. In the case (a) the voltage will be 48V and the capacity 50Ah, in (b) 12V and 200Ah, and (c) 24V and 100Ah, assuming an average 12V unit battery of 50Ah capacity. The Wh capacity will be 2,400 in all cases

been disconnected. A later French inventor, Faure, speeded up the process and produced a lighter battery by coating lead sheets with a paste of lead dioxide. This system is still used in car and truck batteries, to save weight, and such pasted-plate batteries are often called Faure batteries in the industry.

Whatever the system of manufacture, all lead/acid batteries give about 2V per pair of lead plates, or cell. The voltage may vary from about 2.2V when the battery is freshly charged, to 1.7V when it is getting towards flat, but on average 2V is a convenient figure to take. To get higher voltages it is necessary to connect up the cells in series (*see Fig 28*), and a number of cells connected together in a suitable box is what is properly called a battery. The most common arrangements are three cells, giving 6V, used in motor-cycle batteries and also for some heavy-transport systems, six cells (12V) used in car batteries, and various high-voltage systems used for large-scale storage, such as the batteries of 113 cells used for storing mains-voltage electricity.

Apart from the voltage, the most important thing to know about a battery is the amount of power it will hold, its capacity. This is measured in *amp-hours* (Ah) or *watt-hours* (Wh). A good-quality car battery, of the sort fitted to a car of two litres or more, will usually hold about 50Ah, which means that it will deliver an average current of 2½A for 20 hours, or 1A for 50 hours. Theoretically it should also be able to give 20A for 2½ hours or 50A for 1 hour, but in practice high rates of discharge seem to waste some of the energy stored in the battery, so that a typical 50Ah battery may only in fact last for 1.7 hours at 20A or as little as 33 minutes at 50A discharge. For this reason batteries are often quoted as having a certain Ah capacity for a fixed, time of discharge, usually 20 hours. A battery rated at 60Ah if discharged for 20 hours can be guaranteed to give 3A for 20 hours or 1½A for 40 hours (lower rates of discharge do not waste energy), but will not necessarily give 6A for 10 hours.

Watt-hours obviously depend on the voltage of the battery. Our 50Ah car battery, assuming the usual 12V, will give 600Wh, but for the reasons stated it cannot be relied upon to give this power at high rates of discharge.

The batteries most readily available are car batteries, usually 45–55Ah for 20 hours, or 540–660Wh for 20 hours, and these are by far the cheapest batteries available in terms of price per watt-hour (*see Table 8*), because they are produced in such enormous numbers. Other batteries that may be available easily are heavy commercial and passenger vehicle types, used for starting large diesel engines. These are usually 24V, often arranged as four separate 6V batteries in series, and have a total capacity of about 90–240Ah (2160–5760Wh). The heavier American cars tend to use 12V batteries with 70–75Ah capacity, so if these are available to you they have advantages for storage. Traction batteries, used in milk-floats and fork-lift trucks usually run at 30–72V, according to type, and can have capacities up to 10,000Wh. They are expensive to buy new, but may be findable secondhand, and the batteries have a long life if they have been properly maintained. Remember that traction batteries weigh about 9.2lb (4.2kg) per 100Wh capacity, so a 10,000Wh battery will weigh nearly half a ton and occupy about 6½ cubic feet.

Even larger are the stationary batteries specifically intended to be used for static storage of electricity. Some of these are made with capacities of 180,000Wh, and they are extremely long-lasting. However, you are not likely to find them on the secondhand market, and new ones are costly.

If you want a large battery for storage, the cheapest way (unless you are lucky enough to find some large traction or stationary battery going cheap) is to connect up a number of car batteries in series-parallel (*see Fig 28*). If you connect all the positive poles of your batteries together, and all the negative poles, you will have a battery still giving 12V, but the capacity will be the sum of the capacities of the individual batteries. If your power source can cope with a higher voltage, you can also connect the batteries in series in twos or threes to make a larger 24V or 36V battery. Again the total watt-hour capacity will be the sum of all the Wh figures for the batteries.

Suppose, for example, that you are driving a heavy truck alternator and can get about 30V at an average of 60A. The 1800W this alternator can provide will furnish 43,000Wh per day, and if you decide to store half of this for the 'peak' periods you will need batteries with a capacity of 21,500Wh. This could be done with a traction battery or similar large device, but twenty-six heavy-duty car batteries connected together in thirteen pairs (*Fig 28*) will do just as well.

The advantage of using a higher voltage such as 24V is that you need thinner cables, which are cheaper. As we saw in Chapter 2, watts = current in amps × voltage, so if the voltage is higher the current can be less for the same amount of power. However, it is current density that decides the thickness of the cable you need, so if you can run your batteries at 24, 36, 48 or even higher voltages you can save on wiring costs. You must of course have a source of power that produces enough voltage to overcome the natural reverse voltage of the batteries: if your batteries produce 24V, you will need to feed them with at least 27V, if 36V, the input must be 40V, and so on. You will find ideas for cut-outs to deal with variable input voltages on p 108.

Pasted-plate batteries, such as car batteries, have one major disadvantage over stationary batteries, which are usually the Planté type, made with solid sheets of lead. After a lot of charge/discharge cycles the paste of lead dioxide tends to fall off the plates (you can sometimes see it as a sludge in the bottom of an old battery if you break it open), and this lowers the capacity. The Planté battery is made with heavy sheets of lead that form their own lead dioxide each time they are charged, and therefore can be used almost indefinitely, certainly for more that 20,000 charge/discharge cycles. The heavy masses of lead in a Planté battery make it unpopular for carrying on a vehicle, but its working life is greater than that of the pasted-plate battery. If the present energy crisis makes more people install private storage systems, manufacturers may find it worth while to mass-produce Planté batteries as they do car batteries, and this should lower the cost per watt-hour.

All types of lead/acid battery become less efficient when they are cold: a battery with a nominal capacity of 50Ah may be able to produce only 35Ah at freezing point (32°F, 0°C), and in really cold conditions (0°F, −18°C) the capacity will be only about 20Ah. While it is useful to be able to keep your batteries outside the house, you should make sure that they are insulated against the cold: a box lined with rock wool or glass wool may help, or one of the many lagging systems designed for cold-water tanks if the battery is in a rectangular box. There is not much danger of the batteries actually freezing and cracking, sulphuric acid acting as an anti-freeze, unless they are left nearly flat in cold weather, with the sulphuric acid level very low. In any case, as will be considered later, batteries should not be left flat for long periods.

The care and maintenance of lead/acid batteries is fairly simple. One of the most important things to know is the state of the charge. It is possible to assess this by measuring the voltage across the battery—as has been said, a fully charged battery gives about 2.2V per cell and an almost flat one 1.7V per cell, so a nominal 12V battery will go from 13.2V to 10.2V. However, this is not as simple a measurement as it seems, because the load on the battery makes a difference to the voltage, and you do not want to have to disconnect everything every time you measure the voltage.

The simpler method is based on the fact that as a battery discharges the sulphuric acid is converted to lead sulphate. Sulphuric acid is denser than water, so the battery fluid is more dense when the battery is fully charged, and becomes less dense as it discharges. Pure sulphuric acid has a density of 1.8305, water of 1.0000. The acids used for batteries are intermediate, usually with a density 1,270–1.285 (corresponding to about 38–40 per cent sulphuric acid by weight) for car batteries and 1.210 (31 per cent sulphuric acid) for stationary batteries. As the batteries discharge this density gradually falls until it is around 1.110 (16 per cent sulphuric acid).

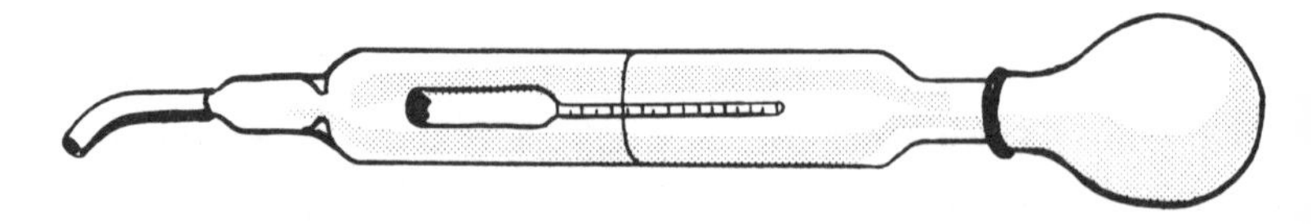

29 Hydrometer for measuring the state of charge of batteries. Acid is sucked into the outer tube by the rubber bulb, and the hydrometer floats to a certain mark on its stem. Normally a fully charged battery gives a reading of 1.26–1.28, a half-charged one 1.200–1.220, and a flat battery 1.110–1.130

The simplest way to measure the density is with a hydrometer, a small floating gauge marked with the densities (*see Fig 29*). For use with batteries the hydrometer can be enclosed in a special syringe with a rubber bulb and tube for sucking a small amount of acid out of the battery cells.

There is one minor complication in this measurement: densities alter with temperature, becoming higher as the temperature falls. If your batteries are very cold or very warm (as, for example, in the depths of winter or if your installation is in the tropics), you will have to apply a correction to your density readings to decide whether the batteries are charged or discharged.

A simple rule for this is to measure the density and then add 0.004 for every 10°F above 60°F, or subtract 0.004 for every 10°F below 60°F (or, in centigrade, add 0.0036 for every 5°C above 15°C and subtract 0.0036 for every 5°C below 15°C). For example, if the weather is freezing, around 30°F, and your hydrometer reads 1.223, subtract $3 \times 0.004 = 0.012$ to allow for the 30° difference in temperature between freezing and 60°. The true state of charge is as if the density were 1.211.

Water has to be added to the battery acid from time to time for two reasons. One is that evaporation can take place, especially in warm weather. The second is that, once the battery is fully charged, any further input of electricity tends to split up the water into hydrogen and oxygen, which appear as bubbles on the plates ('gassing'). If this gassing goes on for a long time a good deal of water may be used up and must be replaced, as apart from anything else parts of the battery plates may no longer be under the acid and therefore cannot operate properly.

The water used to replace these losses must be as pure as possible: distilled or deionised water is best. There is some controversy about this: many people claim that they have used tap water in their batteries for years without any bad effects. It really depends on the water supply. Some cities, Glasgow for example, have water that comes off granite hills and contains almost no impurities except a small amount of peat extract. But in parts of Gloucestershire, for instance, the water contains so much dissolved limestone that it would gradually neutralise the battery acid and make it useless. This is an insidious process because the density of the acid may not change although it is being rendered ineffective. Water from a water-softener is equally bad because it often contains sodium bicarbonate that can also neutralise the battery acid. Unless you know precisely the composition of your tap water, use distilled

or deionised water—its trifling cost is nothing compared t
that of replacing a ruined set of batteries.

Lead/acid batteries should never be charged or discharge
too rapidly. If you short-circuit a battery (for instance b
dropping a screwdriver across the terminals between positiv
and negative), the electric current inside the battery will b
so great that the plates heat up: those in a car battery wil
usually shed most of their lead dioxide paste and lose thei
capacity, but in all lead/acid batteries they will tend to ben
and could easily touch one another, thus permanently short
circuiting the system. Similar damage can occur if the batter
is charged at too high a rate: this is not likely to happen wit
home-made generation systems unless you have grossly under
rated the amount of power produced.

Other sources of trouble with lead/acid batteries may b
summarised as follows:

Sulphation occurs if a battery is left flat for too long, wit
the acid at a density of 1.15 or less. Normally the lead sulphat
produced as a battery discharges decomposes smoothly back int
lead and lead dioxide during the charging process, but in
very flat battery it gradually forms hard crystals that do no
redissolve. One way of bringing a sulphated battery back t
life is to give it a very long charge with a small amount o
current, but the best policy is not to let sulphation occur: if th
storage system is to be shut down for any period, always mak
sure that the batteries are fully charged, if necessary from th
mains.

Oxidation occurs if the battery is charged beyond its capacity
the continued gassing of oxygen bubbles begins to affect th
plates and they become less effective. When batteries are full
charged, either cut off the incoming supply, or set up a bigg
installation to use the extra power.

Gas explosions are not a fault of the batteries in themselve
but of the installation. When batteries are charged there
always a certain amount of gassing, and a mixture of hydroge
and oxygen is given off (this is why batteries have ventin
holes in the caps). This mixture of gases is explosive, and mu
be allowed to escape from any enclosure round the batterie
Never examine batteries with a naked light.

Corrosion occurs at the tops of batteries, and the terminals grow a whiskery deposit of lead salts. This can interfere with the electrical contact and cause a fall in power. Clean the connectors and the terminals thoroughly with fine emery paper, make the connection, and then smear terminal and connector with petroleum jelly. Make sure that none of the jelly or grease gets in between the terminal and connector, for it will act as an insulator.

Spillage of acid must be watched. The sulphuric acid used in batteries is quite strong enough to damage skin, clothes, and surfaces. (For acid on clothes or skin, wash off immediately with plenty of water; for spills in awkward places, sprinkle washing-soda crystals freely around until it is possible to wash the area down.)

If you have to replace acid spilled from batteries, you can make up acid at density 1.280 by adding 1 volume of concentrated sulphuric acid to 2.9 volumes of distilled water. Always add the acid to the water: the mixture gets very hot, and if you pour water into the acid it may be converted to steam and splash over you.

Alkaline batteries

There are two types of alkaline battery in general use, the nickel/iron battery and the nickel/cadmium battery. The active plates are made of the metals specified, and instead of sulphuric acid they contain an alkaline solution of potassium hydroxide ('caustic potash') and lithium hydroxide at about twenty per cent by weight—the density recorded by the hydrometer is between 1.17 and 1.19. The chemistry of these batteries is very complex and is not entirely understood even now, despite the fact that Edison invented the nickel/iron battery in 1900.

For many purposes, including storage, alkaline batteries have several advantages over the lead/acid type:

1. They are lighter in weight for a given capacity.
2. They are mechanically stronger and can stand vibration or shock without losing capacity.
3. They have a service life at least four times greater than a pasted-plate lead/acid battery.

4. They are not damaged by rapid charging or discharging.
5. They can be left flat for long periods without deterioration.
6. They do not lose charge if left for long periods—this makes them very useful for emergency supplies.
7. They need less maintenance—the only normal service necessary is topping-up with distilled water.
8. They are not as affected by cold weather as lead/acid batteries.

Alkaline batteries have two major disadvantages: they give only about 1.2V per cell, which means that more cells are necessary for a given battery (20 cells for a 24V battery instead of the 12 cells for a lead/acid type), and they are more expensive. Even on the secondhand market a 550Wh aircraft-engine starting battery will cost at least £50, and this is only equivalent to one car battery. On the other hand, they will last much longer, and can safely be bought secondhand because even abuse does them little harm.

One minor difficulty is that it is difficult to tell the state of charge. The alkaline solution does not change density like the acid in a lead/acid battery, and the limitations on voltage measurement found with other storage batteries still apply here, especially as alkaline batteries do not drop off in voltage until almost flat. For a storage installation fit an ampere-hour meter which actually measures the output.

When charging, alkaline batteries produce explosive mixtures of hydrogen and oxygen, and the same precautions should be taken as with lead/acid batteries. The caustic potash solution is very corrosive to skin and clothes, and should be washed off immediately with plenty of water if spilled. For a large spill, use acid or lavatory cleansing powder to neutralise the alkali.

Silver/Zinc batteries

These are the lightest storage batteries per unit of capacity at present on the market, and their design has had a great boost from the needs of spacecraft. An 85Ah 12V Venner silver/zinc battery weighs only 17lb (7.7kg) compared with the 68lb or so for a similar lead/acid battery. They are still much more expensive than lead/acid or alkaline batteries, and also have a shorter life in terms of the number of charge/discharge cycles they can take. Both these disadvantages may be overcome in a relatively short time.

Comparison of storage batteries

Table 8 shows comparative figures for various types of storage battery. The figures for £/Wh can only be approximate as there are so many price variations within each class, and therefore the figures of £/Wh/cycle must be similarly approximate. They do however give a rough idea of the cost of setting up and maintaining a storage system at any given capacity. For example, it is clear that, despite the higher capital outlay, stationary lead/acid batteries of the Planté type are still the cheapest storage system, followed by traction batteries, nickel/iron batteries and ordinary lead/acid car batteries. The last, however, give the lowest initial cost.

Table 8 Comparison of Storage Battery Types

	Lead/acid types					
	Car batteries	*Traction batteries*	*Planté batteries*	*Nickel/cadmium*	*Nickel/iron*	*Silver/zinc*
Open circuit (maximum) voltage	2.10	2.12	2.06	1.3	1.3	1.86
Normal operating voltage	1.98	1.94	1.94	1.2	1.2	1.45
End of charge voltage	2.53	2.55	2.17	1.48	1.48	2.05
Recommended working temperature °F (°C)	70–90 (21–32)	70–110 (21–43)	70–90 (21–32)	65–85 (18–29)	65–90 (18–32)	50–90 (10–32)
Weight (lb/100Wh capacity)	7–9	10–11	20–25	5–7	5–7	2–3
Cycle life	150–250	1000–2000	10,000	10,000+	10,000+	25–50
£/100Wh	0.6–1.2	2–4	5–27	25–100	20–80	50–80
£/100Wh/cycle	0.002–0.008	0.0015–0.002	0.0005–0.0027	0.0012–0.01	0.001–0.008	1–3.2

Other developments

Because of the great interest in alternative sources of electricity, and also the need to produce new vehicle systems not dependent on petrol or diesel fuel, much research goes on into new storage-battery systems, particularly batteries that can be made cheaper and lighter than the traditional lead/acid type.

A storage battery favoured for driving electric vehicles is the sodium/sulphur battery. The metal sodium and the non-metal sulphur are both much lighter than lead or nickel, and can therefore hold far more electricity per unit weight of battery. However, there are two disadvantages. Sodium reacts with water vigorously, so the battery has to be made with a special ceramic material instead of the usual acid or alkali solution. If the battery is crushed, in a vehicle accident for example, and swamped with water, it can become dangerously hot because of the reaction of the sodium with the water.

The other disadvantage is that sulphur is solid at room temperature, and will only work in a battery if it is kept melted, so the battery has to be kept at about 250–300°C (482–572°F). Fortunately the temperature rises as the battery is charged, so if the whole system is kept in an insulating box, only small amounts of extra heat have to be added. Usually the batteries are made with a small heating coil that takes some of the charging current to keep the cells hot.

Although the sodium/sulphur battery was primarily designed for use in electric vehicles, if installed to store electricity in a building there would be very little danger of damage to its cells, and keeping it hot would be simple.

The sodium/sulphur single cell gives about 2.05V open-circuit, and 1.75V under load (about the same as a lead/acid battery cell, in fact), but the main difference between the two types may be summarised by the fact that a 50,000Wh sodium/sulphur battery weighs about 1760lb (800kg): a lead/acid battery of similar capacity would weigh 5,000lb (2272kg) and the Planté type of battery would weigh even more.

Other new storage batteries are the lithium/sulphur and lithium/chlorine types. Both of these are again much lighter than lead/acid batteries of similar capacity: but so far the difficulties in mass-producing them efficiently are greater than with the sodium/sulphur battery.

Connections to batteries

As said in Chapter 2, in low-voltage installations the currents carried by the cables are far higher than in ordinary domestic wiring, and the cables must therefore be much heavier. This is particularly so with the connections to a system of batteries, as all the current for the whole installation has to run through them.

Connectors between batteries which are made up into series or series/parallel arrays should be heavy lead-coated copper or similar metal straps, bolted firmly on to the battery terminals, and coated with petroleum jelly to avoid corrosion. The lead-out wires should be able to carry the whole current load for the system (see Appendix 1 for suitable wire dimensions) with no more than 2.5 per cent loss of voltage.

11 Inverters and Converters

While it is often useful or even essential to store your electricity supply in the form of a low-voltage dc output from batteries, it would often be convenient to have a supply at standard mains voltage ac in order to run ordinary household equipment. While 12V lights, radios, TV sets and so on are available, they often cost more than equivalent apparatus made for ordinary mains voltage, because there is less demand. Many small pieces of mains-driven equipment that add to the comfort of life are just not available in 12V modifications, or very hard to find.

Also, if your main purpose in building an alternative electricity supply is to have insurance against power failures, strikes and shortages, you cannot duplicate every essential piece of electrical equipment. You need to be able to switch over your deep-freeze, central-heating pump, and other essential equipment quickly.

For these reasons it is usually necessary to have some means of changing at least some of your low-voltage dc supply to 240V ac (or whatever your public supply may be). Equipment that changes the voltage of a supply from low to high or vice versa is usually called a converter: in the special case where dc is changed to ac it is called an inverter. (There is some confusion in the electrical world about these words, and you may find all types of supply modifier called converters, but the above definitions seem less confusing).

Ac's great advantage over dc, which has made it the almost universal type of supply, is that ac voltages can be changed very easily and with little loss by a transformer, while dc voltages cannot. Once you have converted dc to ac at any voltage, therefore, it is easy to produce a supply at 240V or any other required voltage.

Rotary machines

The simplest way to change dc to ac, in theory, is to use the dc to run a motor which drives an ac generator. This is the principle of the rotary converter, which can be used to produce dc from ac (often done in factories where dc is required for electric motors, plating, welding, and other specialised requirements), or ac from dc (the inverted mode). The equipment is also often called a dynamotor, although properly this should be reserved for the type of converter that makes high-voltage dc from low-voltage dc.

Rotary converters can often be found on the secondhand market, and they are quite a good method of producing ac from 12V dc batteries. Make sure when you buy one that it will work from dc to ac. With dc motors it does not usually matter whether you are feeding electric power in and running them as motors, or turning the shaft by some external means and running them as generators—they are reversible. On the other hand, many ac motors, especially the induction motors used for smaller loads, are not reversible, and will not give any ac output if you turn them by some external drive.

The electrical layout of a rotary converter may look a little complicated, but it is really just two machines in one case and on one shaft. There is no electrical connection between the motor and the generator. Even though the ac generator needs a dc supply for its field coils, this is usually kept quite separate from the field windings of the dc motor.

The intrinsic faults of the rotary converter are that it is inclined to be noisy, because of the moving parts, and there is a considerable loss of power due to losses at the armature of the motor and the frictional and similar mechanical losses throughout the mechanism.

A secondhand converter can be serviced by simple first-aid such as cleaning the armature and slip rings, replacing brushes if they are worn, and generally cleaning up the bearings and other mechanical parts. If a converter overheats it is usually because of a short-circuit in the windings, and this will lower the efficiency of the machine as well as being a fire hazard.

'Hunting' or a fluctuating rise and fall in output may be due to serious faults like a breakdown of the voltage-regulating system, but it can be caused by poor contact between the slip rings and brushes if the alternator is of the rotating-field type, and simple cleaning will probably put things right.

Vibrating machines

Another way of making a dc supply discontinuous, so that it can be stepped up by a transformer, is to use the dc power to operate a vibrating contact which connects and cuts off the dc power at the desired frequency. Such a device is used in the common type of electric bell running from dc batteries, where the vibrator is also the hammer that hits the bell, and it used to be familiar in the 'induction coils' employed to generate very high voltages from a low-voltage dc supply. When it was believed that shocks from high-voltage electricity were somehow good for the health, the induction machines were familiarly called 'medical coils', and old ones may still be encountered in junk shops.

Vibrators have been almost entirely superseded by electronic oscillator circuits with no moving parts.

Oscillator inverters

There are many ways of making electronic circuits oscillate, so that a steady dc voltage can be converted to an ac type of output. Such circuits are essential for the operation of radio, TV, radar and other electronic equipment. Almost any of these oscillator systems could be adapted for inverting dc power to ac, ready for transforming to 240V, but it must be remembered that the currents flowing in such circuits as radios are very small compared with the heavy currents in a power supply. Power inverters must use components, especially the transistors and other semiconductors, which have high wattage ratings, otherwise the components will quickly burn out or be destroyed by the heat generated as the current flows. Such components are quite expensive, so that designs using the minimum number of semiconductors are preferred for power supplies.

The simplest way to create ac is to apply the dc current to two 'gate' components, and arrange for the gates to be opened alternately, 50 times per second for a 50Hz supply (*see Fig 30a*).

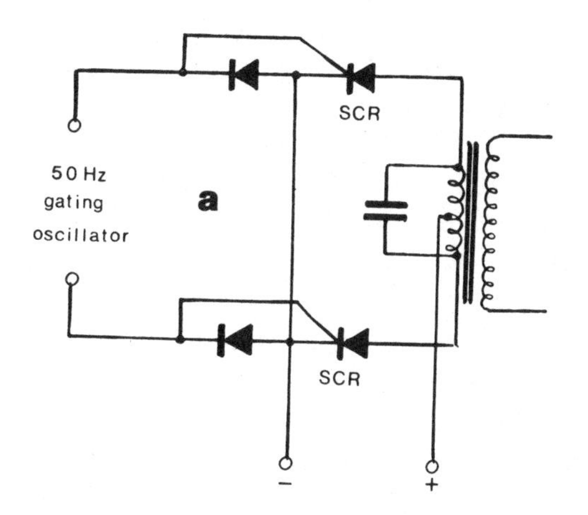

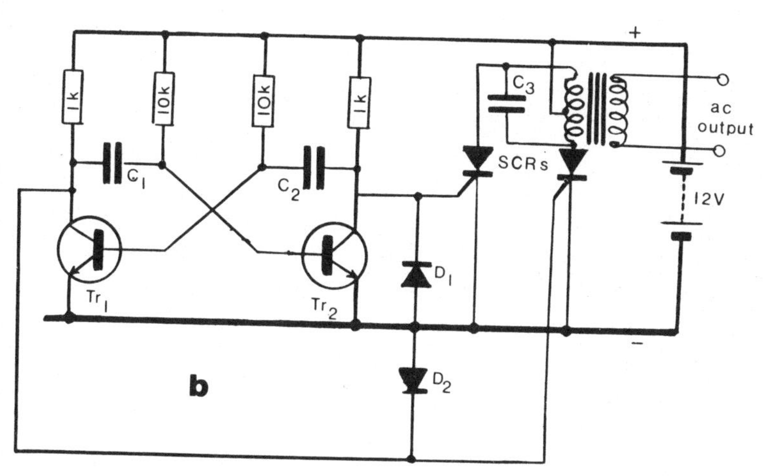

30 Inverter based on silicon-controlled rectifiers (SCRs): (a) principles of circuit, (b) practical circuit. The transistors Tr_1 and Tr_2 are BC109 or similar types, C_1 and C_2 1.4μF, and C_3 0.1μF

Here the gates are two thyristors or *SCRs* (silicon-controlled regulators). These devices pass current only when a small voltage is fed to their gate terminals: when this happens they switch on and can pass quite large amounts of power, some up to about 18kW. For the more moderate wattages used in home power supplies there are many suitable SCRs available. The current through the SCRs oscillates through the transformer

(T) primary coil at the same frequency as the gating unit, and by proper choice of the gating unit and the numbers of turns in the primary and secondary coils of the transformer 240V at 50Hz can be taken off to the load.

Figure 30b shows a practical circuit in which the gating unit is a simple type of astable multivibrator running at 50Hz. This can easily be modified by selection of other SCRs if they are available more cheaply or easily.

While this system is relatively simple and robust, it has one practical disadvantage. The output is not a smooth ac supply, but a square-wave supply that leaps suddenly, almost instantaneously, to its maximum positive voltage, then as suddenly drops to its maximum negative voltage. This has several effects. Radio and TV interference may result from radiations set off when the sudden voltage changes take place, so it is more difficult to reduce 'mains hum' when using the unit, and because of the sudden transitions in voltage, induction motors are not able to take advantage of all the power, so that central-heating pumps, refrigerators of the compression type, and deep-freezes may not run as efficiently as with the smoother mains ac. This may not matter unless the motors get rid of the unused power by heating up: that depends to a certain extent on the design of motor, but it is as well to check when using this type of inverter for the first time.

Another consequence of the square-wave characteristic is that the supply stays at its maximum voltage value for a longer time than the more smoothly oscillating sinusoidal mains ac. This means effectively that more watts go in for a given voltage, and this can affect equipment such as TV and radio sets where there are valve filaments to be driven by the supply (even in a transistorised TV the picture tube still has a filament). These filaments are designed to run on a very accurately measured supply, and if you put more watts through them than they are designed to take they may burn out or give an unsatisfactory performance in other ways. If, therefore, you want to run your TV or mains radio from a supply which has square-wave or mains radio from a supply which has square-wave characteristics, set the voltage adjustment control at the back of your set to a higher value than the nominal voltage of your supply: if the supply gives a reading of, say, 220V when connected to the set, put the voltage control to 230 or 240V to be on the safe side.

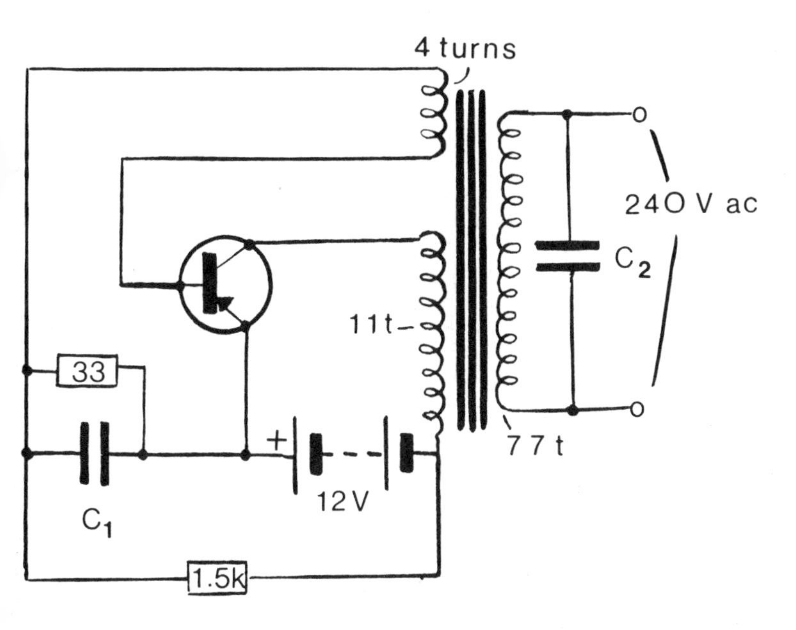

31 Sine-wave inverter for small loads. The transistor is a 2N 2147 or similar, and C_1 8μF. The transformer turns are wound on a ferrite core

It is of course possible to make inverters that give a reasonably smooth sinusoidal output, but these are rather more difficult to design for large wattages. Figures 31 and 32 show two designs, a very simple one-transistor circuit giving about 10–15W maximum, and a more elaborate circuit for about 100W.

Remember when building any of these inverters that the currents in use are greater than in normal electronic circuits, and thin copper connectors as in Veroboard and ordinary printed circuits are not good enough for the parts of the circuit that carry the full power; use thick copper wire. For instance, in the SCR inverter in Fig 30b, it is satisfactory to build the gating unit on Veroboard or a printed-circuit board, but wire up the SCRs with heavy wire to carry the amperage (see Appendix 3 for current ratings for various thicknesses of

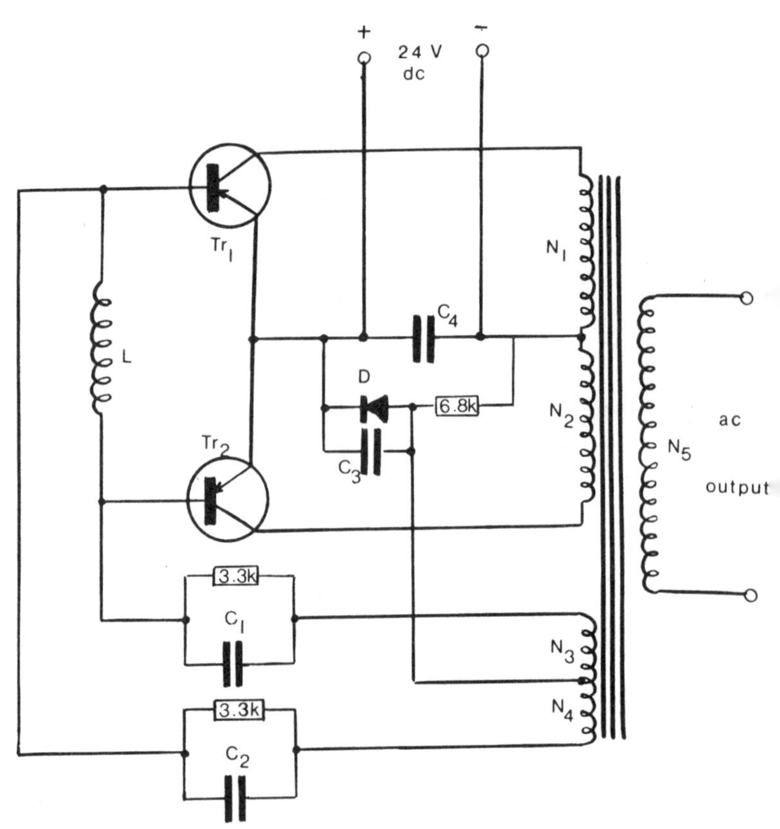

32 Larger sine-wave inverter. Tr_1 and Tr_2 are OC 28 or similar types, C_1 and C_2 50 μF, C_3 0.1μF and C_4 0.2μF. The choke L is made up of 97 turns, N_1 and N_2 130 turns, N_3 and N_4 18 turns, and N_5 1330 turns

wire). The transistors or SCRs must be mounted on heat-sinks of sufficient size to dissipate the heat, otherwise your inverter will fail after a few minutes, when the heat has destroyed the transistors. For the circuits in Fig 31 and 32, for instance, mount the power transistors on pieces of aluminium at least 6 square inches in area and held upright to help to circulate the air. For any larger unit it would be better to think in terms of mounting the power transistors on SCRs on a finned heat-sink fixed to the outside of the case of the unit, or to incorporate one of the small fans that are now available for ventilating such equipment.

Inverters are now available ready-made from a number of suppliers, and a study of the trade papers will, if you are interested in buying a complete unit, give you a choice of capacities in watts and prices. Most of the commercial inverters tend to be based on square-wave circuits, and therefore need the care in use described earlier in this chapter.

12 Changeover switches

If your alternative electricity supply is to be used mainly for emergencies, and you will be using the ordinary mains supply most of the time, you will find it useful to fit up some system to turn on at least some of the emergency power automatically as soon as the mains supply fails. Otherwise, if you are asleep or not in the house when the power cut comes, you may not even know about the loss of power until serious damage has been done.

If you have the more expensive sort of petrol- or diesel-driven generator which has a self-starter, there is no problem. Most of these have automatic changeover switches to start them as soon as the mains supply goes off, and if not, such switches can easily be bought and fitted.

If you have a simpler and less expensive system, you can still give yourself the benefit of automatic changeover. Suppose that your supply is stored in 12V batteries and your emergency supply runs from this battery stack, either at the 12V level or through an inverter. In theory, the simplest plan would be to fit a relay between the battery and the load, the coil of the relay being operated from 240V ac mains, and the contacts normally open when the coil was energised. This would mean that the battery was cut off as long as ac mains passed through the coil, but as soon as the mains failed the contacts would close and connect the battery.

In practice this idea suffers from several disadvantages. First, it is quite difficult to find relays that work from ac mains with sufficiently heavy contacts to carry the large currents of battery-powered circuits. The kind of relay used in cars, for example for the starter-motor, would carry the currents, but the coils of these relays are meant to operate from 12V dc. Second, as set up, the coil of the relay would be energised all the time except during fairly rare times of emergency. This would not consume much power, because relays are not usually heavy power users, but the life of the coil would be limited by the constant current flow, and the contact control could get 'tired' and sluggish. If the coil burned out, not only would you have the replacement of the relay to worry about, but your emergency system would be switched on, perhaps when you were out of the house, and there could be complications if the emergency power and the mains were on at the same time.

Obviously it is better to find a system that both has some safeguard in case the relay burns out and uses one of the existing types of car relay where the coils take 12V dc. This can be done by a simple modification of the circuit that also provides a trickle-charge for the batteries from the mains.

The mains supply is fed into the transformer and full-wave rectifier system (*see Fig 33*) so as to provide a dc supply at about 15V. An ordinary commercial battery charger can be used for this supply, or you can easily fit up the supply with a suitable transformer and metal-plate rectifier. The current supply from this device need only be about ½A, as it is only intended to give a very slight constant charge to the batteries (unless, of course, you have a very large battery set-up, in which case you may wish to give more constant-charge current).

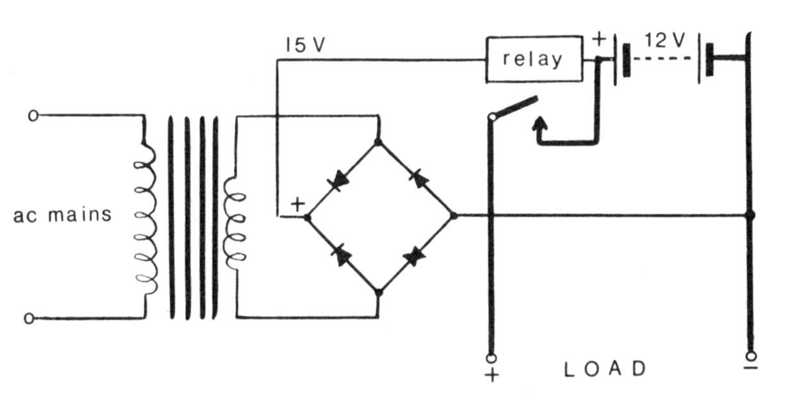

33 Simple relay changeover switch. The relay contacts must be rated for the maximum load (see text for use of car starter motor relay)

A car starter-motor relay is then fitted up as shown (*see Fig 33*) so that its coil runs between the positive line from the battery charger and the battery positive terminal (assuming a negative-earth system). A resistance R is also included to help balance the voltages so that no current runs through the relay coil when the battery charger is actually providing current. The value of this will depend on the precise voltage your rectifier system gives. The resistance may not even be necessary: the idea is just to make sure that the voltage across the relay coil is considerably less than 12V.

The contacts of the relay are now connected to the positive pole of the battery and the cable to the load, remembering to use sufficiently thick wire to carry the maximum current in the system (see Appendix 3 for recommended current ratings).

As soon as the ac mains fail, the battery charger stops providing a voltage, and the full battery 12V is across the relay coil, so that the contacts close and the battery is connected to the load. When the mains come on again, the coil only receives about 2V and the contacts upen again. A trickle-charge flows through the relay coil to the battery, but this very small current is not likely to damage the coil.

If you have a more complex system that really needs personal attention to start it, you can make a warning device by using the same circuit but with a smaller relay, using the supply from the battery to run a light or buzzer which goes on as soon as the ac mains fails.

There are better and more complex changeover switches on the market, and no doubt some readers will be able to think of modifications; but this simple circuit will prove quite reliable for a medium-sized emergency system.

13 Safety

Enthusiasts who build electronic equipment at home often suffer from what I call the 'My Baby' syndrome—a belief that, because they have built the piece of equipment with their own loving hands, it could not possibly hurt them. This is probably true with battery-driven circuits, but equipment driven from the mains has only too often proved that it has no sense of filial affection. When you consider that radio transmitters, TV sets and photomultipliers often use voltages measured in thousands, it is not really surprising that some unpleasant accidents occur.

The same thing can happen if you are building your own power supply: it is somehow difficult to believe that the power coming out of your wind generator, home-made engine or inverter is as 'real' as the electricity that comes from the mains. Indeed, the power available from a home-made generator is usually far less than that supplied to the consumer unit of the mains supply; but as a current of ½A passing through the body is usually fatal, it is not wise to depend too much on the limitations of the home generator.

The severity of an electric shock depends on these factors:

1. The amount of current that passes through the body. It is current that injures, not voltage. Many people have had a shock of some thousands of volts from a static-electricity machine such as a van der Graaf generator or Wimshurst machine, without much more reaction than momentary discomfort, because the amount of current supplied by these machines is infinitesimal. On the other hand a heavy current

at 110V can kill, and there have been deaths from voltages as low as 46V.

A current of 20–30mA applied suddenly and not continued will give an unpleasant shock, but causes little permanent damage (unless, of course, you are standing at the top of a ladder when you get the shock). Currents of 70–100mA will cause some permanent damage and burning of the skin which has been in contact with the supply. Currents of 500mA are nearly always fatal.

2. The resistance of the skin affects the amount of current that can penetrate at any given voltage. Dry, normal skin has a resistance of about 50kΩ/cm², which means that if you touch a 240V live plate with your hand, and 10cm² of your skin is in contact, you will receive a current of 48mA, enough to give you a nasty shock but probably not seriously damaging. On the other hand, sweating can lower the resistance by as much as twelve times, which could mean, in similar circumstances to those above, a current of 570mA, which would almost certainly be fatal. Actual water on the skin can lower the resistance as much as twenty-five times, which is why electric shocks sustained in the bath or similar wet places are so serious, and why regulations forbid the placing of power points in bathrooms.

Once inside the skin, electricity travels easily, as the blood vessels conduct quite well; a shock often goes straight to the heart.

3. The path of the shock through the body makes a lot of difference to the kind of damage. If you put your fingers into a light socket by accident, the current will probably pass from one finger to the other, burning and hurting you, but probably not doing much more harm. If you touch a live conductor with one hand and an earthed water-pipe with the other, the current will pass from arm to arm via the chest, and your heart or lungs may be affected. Currents passing through the head often cause permanent nervous damage.

4. The length of time during which the current is applied is important. As the current passes it lowers the resistance of the skin, so that the effects get worse. Unfortunately, one of the usual effects of electric shock is to 'freeze' the muscles, so that it

is often difficult to let go of a live wire or tool until the current is actually turned off.

5. The type of electricity also affects the amount of damage sustained. An ac supply gives a worse shock than dc of the same voltage and current; but on the other hand if a dc supply starts arcing to the body it goes on longer before the arc breaks down.

What to do in case of shock

If you get a bad shock and can move at all, try to reach the switch or cut off the supply somehow. This is easier than trying to pull yourself free from the live supply. If you get the kind of shock where you are thrown across the room, rest for a while and have a hot sweet drink. Look carefully at your hand, or whatever other part of you touched the live conductor, and if there is a bad burn get it treated by a doctor or hospital emergency department. Burns from electric shock can go a lot deeper than appears from the outside. Experience in industry has taught me that electric shocks can have unpleasant long-term effects if not properly cared for.

If you see someone else obviously 'frozen' to a live conductor, do not try to pull him or her off with your bare hands. The shock will simply spread to you and you may find yourself equally helpless. Try to switch off the current or otherwise break the circuit (pull out a fuse, unplug a flex), but if you cannot immediately see to do this, push the victim away from the electricity supply with a piece of wood, a chair (not a metal-framed one), or any handy non-metal object. If the person is unconscious, use mouth-to-mouth respiration and call a doctor. Shocks can often temporarily paralyse the muscles of the chest, and stop breathing.

Avoiding shock

A few simple precautions can ensure that you never get even a mild shock from any of your projects.

1. When working on any piece of equipment with voltages over about 50V, always disconnect it from the supply. Do not rely on switches, which can fail or be wrongly connected. This is particularly important when working outdoors, because

water can have got into switches or other isolating devices, and your own skin will be less resistant because of damp.

2. If you have to work on a supply in emergency (if, for example, the power line from your wind generator has come down and there is no easy way of stopping the generator at once), wear rubber gloves or even the disposable polythene gloves sold for household work, especially if you are working at 50V or more.

3. When using high-voltage supplies as from an inverter or 240V generator, check any metal casings on equipment from time to time. If you have, for example, mounted your inverter in an aluminium case, check occasionally with a multimeter or voltmeter that a leakage has not occurred between the high-voltage side and the metal casing. If the casing is insulated from earth you may not notice anything wrong until you happen to touch it with one hand while holding an earthed conductor with the other.

4. For the sake of your equipment as well as yourself, always fit suitably-rated fuses in the circuit as near to the generator or inverter as possible. You can easily get fuse holders from a scrap-car yard, and the appropriate heavy-duty fuses for low-voltage supplies are available as standard car spares. If you decide to use contact-breakers instead of fuses, make sure that they are in good condition and not corroded by weather or old age: remember that a heavy current can weld metal parts together, so once your contact-breaker has stuck it is likely to go on passing a short-circuit current until something else catches fire—usually your generator.

5. Do not try to save money by using cables too thin for the currents you are passing. The temptation is great, especially if you happen to have supplies of domestic-style cable, but for low-voltage work heavy cable is absolutely necessary because of the large amperage. Apart from the danger of fire you will lose a great deal of your electricity if the cables are unsuitable.

6. Remember that the ratings for most cables are based on the assumption that they are properly ventilated, so that the warmth produced by the current passing is dissipated. If cable is kept in a large coil, or otherwise prevented from dissipating its heat, it may warm up dangerously even though the current load-

ing is well within its specified rating. As an instance, someone wishing to move his television set across the room from the nearest power point used a 100ft extension cable. Not wishing to cut the cable, he coiled the surplus neatly under the carpet. Although the cable was rated for 5A, and the TV set used only about 2A, he found one day that the cable was busily burning its way through the carpet. The first danger sign, he realised later, was that his cat had taken to sleeping on the carpet where the coil was sited—evidence of a warm patch! The combination of a large number of turns of cable close together, a carpet preventing ventilation, and a cat as a final heat barrier, meant that heat was concentrated to many times the level envisaged by the engineers who decided on cable ratings.

7. If you have overhead cables running from an outside generator, use extra supports if there is more than a 10ft (3m) span, and try to keep the cable at least 12ft from the ground, preferably more. You can use ordinary PVC-covered cable for overhead lines quite safely, but make sure that the cable cannot swing in the wind too much, or the copper wires inside may crack apart at points of particular stress. A good system for support is to hang a piece of strong galvanised wire rope between the points of suspension, and attach your cable to this with short lengths of nylon cord. This takes the strain off the cable.

8. If you have small children in the house, make sure that your generator and all the associated electrical equipment is safely shut away in a box or shed that you can lock. Set up your equipment so that neither you nor members of your family and casual visitors can possibly come into contact with high-voltage live parts. Use the same standards as you would if you were setting up an installation for someone else with no knowledge of electricity.

9. Take equal care with the mechanical parts. The rotating shaft of an engine driving a generator can easily catch your sleeve, tie, or any other trailing piece of clothing and pull you towards the moving machinery. A belt running over a pulley can trap your fingers, sometimes with serious consequences. No elaborate fencing off is needed: a simple shield made of plastic or plywood in front of moving parts is as good as an

elaborate metal guard. If the hazard is a rotating shaft, you can often enclose it with a piece of plastic water-pipe held on a wooden support.

As said in Chapter 6, wind-driven generators can get up to very high speeds, particularly the propeller type, and a blow from a blade could cause serious injury. Make sure that such rotors are mounted high enough to clear the heads of passers-by: this will also tend to improve performance, the air flow being better ten or so feet above the ground. Try to position such propellers so that if the blades should break in a high wind the pieces will not go through someone's window or across the public road.

Make sure that you have a reliable method for stopping a wind generator, particularly a high-speed type, or at least some means of turning it 'out of the wind' if the wind-speed gets to a dangerous level.

10. If you use storage batteries, be careful of the sulphuric acid in lead/acid batteries and the caustic potash in nickel/iron and nickel/cadmium batteries. Both are extremely corrosive. Sulphuric acid will attack your skin, most types of textiles and metals, and it makes black stains on wood and plastics. Wash off immediately (*see page 95*).

Caustic potash is particularly corrosive to skin and to such textiles as wool; again wash it off immediately. Spills on other surfaces should be washed with plenty of water; spills in awkward places can be neutralised with vinegar or a small amount of an acid cleanser, such as Harpic.

If you need to refill a lead/acid battery with sulphuric acid for any reason, always add the acid to water to dilute it, never the other way round. Sulphuric acid and water react to produce a great deal of heat, and if you pour water into the acid the first few drops of water may get so hot that they produce a sudden spurt of steam that can shower drops of acid all over you. Wash out any vessels you have used for acid very carefully in running water.

Appendices

Appendix 1 Imperial Standard Wire Gauge

This table shows the Imperial swg numbers followed by the diameter of the wire in inches, the cross-sectional area in mm^2 to relate to the modern metric measurements for wire, and the approximate resistance in ohms per metre for each size of wire when made of copper and in a single conductor (not twisted or braided cables).

swg	*diameter*	*cross-section*	*resistance/metre*
7/0	0.500	126.68	0.000138
6/0	0.464	109.09	0.000161
5/0	0.432	94.56	0.000185
4/0	0.400	81.07	0.000216
3/0	0.372	70.12	0.000250
2/0	0.348	61.36	0.000285
1/0	0.324	53.19	0.000329
1	0.300	45.60	0.000384
2	0.276	38.60	0.000454
3	0.252	32.18	0.000544
4	0.232	27.27	0.000642
5	0.212	22.77	0.000769
6	0.192	18.68	0.000937
7	0.176	15.70	0.00112
8	0.160	12.97	0.00135
9	0.144	10.51	0.00167
10	0.128	8.30	0.00211
11	0.116	6.82	0.00257
12	0.104	5.48	0.00320
13	0.092	4.29	0.00408
14	0.080	3.24	0.00540
15	0.072	2.63	0.00666

swg	*diameter*	*cross-section*	*resistance/metre*
16	0.064	2.07	0.00844
17	0.056	1.59	0.0110
18	0.048	1.17	0.0150
19	0.040	0.811	0.0216
20	0.036	0.657	0.0266
21	0.032	0.519	0.0338
22	0.028	0.397	0.0441
23	0.024	0.292	0.0600
24	0.022	0.245	0.0715
25	0.020	0.203	0.0863
26	0.018	0.164	0.107
27	0.0164	0.136	0.129
28	0.0148	0.111	0.158
29	0.0136	0.094	0.186
30	0.0124	0.078	0.225
31	0.0116	0.068	0.257
32	0.0108	0.059	0.297
33	0.0100	0.051	0.343
34	0.0092	0.043	0.407
35	0.0084	0.036	0.486
36	0.0076	0.029	0.604

Appendix 2 Brown & Sharpe Wire Gauge

For those readers carrying out projects based on US specifications, this Table gives the diameter in inches, the cross-sectional area in mm² and the resistance in ohms per metre for wires expressed in American Wire Gage (AWG, Brown and Sharpe Wire Gauge).

AWG	*diameter*	*cross-section*	*resistance/metre*
6/0	0.58	170.46	0.000102
5/0	0.5165	135.18	0.000129
4/0	0.46	107.22	0.000163
3/0	0.4096	85.01	0.000206
2/0	0.365	67.51	0.000259
1/0	0.3249	53.49	0.000327
1	0.2893	42.41	0.000413
2	0.2576	33.62	0.000520
3	0.2294	26.66	0.000656
4	0.2043	21.15	0.000827
5	0.1819	16.77	0.00104
6	0.1620	13.30	0.00132
7	0.1443	10.55	0.00166
8	0.1285	8.367	0.00209
9	0.1144	6.631	0.00264
10	0.1019	5.261	0.00333
11	0.0907	4.168	0.00420
12	0.0808	3.310	0.00529
13	0.0720	2.627	0.00666
14	0.0641	2.082	0.00840
15	0.0571	1.650	0.0106
16	0.0508	1.308	0.0134
17	0.0453	1.040	0.0168

AWG	*diameter*	*cross-section*	*resistance/metre*
18	0.0403	0.823	0.0213
19	0.0359	0.653	0.0268
20	0.0320	0.519	0.0338
21	0.0285	0.412	0.0425
22	0.0253	0.324	0.0540
23	0.0226	0.259	0.0676
24	0.0201	0.205	0.0854
25	0.0179	0.162	0.108
26	0.0159	0.128	0.137
27	0.0142	0.102	0.172
28	0.0126	0.080	0.219
29	0.0113	0.065	0.269
30	0.0100	0.051	0.343
31	0.0089	0.040	0.437
32	0.0080	0.032	0.547
33	0.0071	0.026	0.673
34	0.0063	0.020	0.875
35	0.0056	0.016	1.094
36	0.0050	0.013	1.346

Appendix 3 Current Ratings for Cable

This Table gives recommended maximum currents (amps) to be carried in twin- or three-core cables made of copper and insulated with PVC or similar plastic. Because of the necessity for adequate heat dissipation (*see p 114*) there are two sets of figures, one for cables enclosed in conduit or trunking, and the other for cables clipped to an open surface. If the cables are at all restricted in terms of ventilation, use the first column to be on the safe side. The cables are defined by the cross-sectional area (mm^2) of the conductors and the number of strands: thus 7/0.85 is a cable made up of 7 strands each with a diameter of 0.85mm.

Cross-section (mm_2)	*Cable type*	*Enclosed cable in conduit or trunking*		*Cable clipped direct to open suraface*	
		Twin	*3-core*	*Twin*	*3-core*
1.0	1/1.13	11	9	12	10
1.5	1/1.38	13	12	15	13
2.5	1/1.78	18	16	21	18
4	7/0.85	24	22	27	24
6	7/1.04	30	27	35	30
10	7/1.35	40	37	48	41
16	7/1.70	53	47	64	54
25	7/2.14	60	53	71	62
35	19/1.53	74	65	87	72
50		—	—	140	125

Appendix 4 Some Suppliers

Engine-driven generators

Centrax Ltd, Shaldon Road, Newton Abbot, Devon TQ12 4SQ (Newton Abbot 2251) Large industrial generators, 500–630kVA

Honda (UK) Ltd, Power Road, Chiswick, London W4 (01–995–9381) Domestic generators, 250W–4kW

Jonlaw Engineering Co Ltd, 37 West Road, Oakham, Leics (Oakham 2261) 1.5kVA–6kVA

Lewis Electric Motors Ltd, Moor Works, Blackamoor Lane, Maidenhead, Berks (Maidenhead 21216) 0.8kW–650kVA, self-starting and automatic

Wind generators

Conservation Tools & Technology Ltd, 143 Maple Road, Surbiton, Surrey KT6 4BH

Ralph Howe Marketing Ltd, New Orchard and High Street, Poole, Dorset (Poole 77377) Complete marine or caravan installation

Parts for wind generators

Glentrade, 60 Balgonie House, Jamphlars Road, Cardenden, Fife (Cardenden 804) Slip rings etc

Joseph Lucas (Sales and Service) Ltd, Great Hampton Street, Birmingham B18 6AU (021–236–5050) DC generators

Solar cells

Ferranti Ltd, Gem Mill, Chadderton, Oldham OL9 8NP (061–624–6661)

Joseph Lucas (Sales and Service) Ltd, Great Hampton Street, Birmingham B18 6AU (021–236–5050)

Inverters

Northern Electro Engineering, Whittle Street Works, Tottington Road, Bury, Lancs

Scientific Products, Onward Building, 326 North Promenade, Blackpool, Lancs

Nickel/cadmium batteries

G. W. George, Mallards, South Hanningfield, Chelmsford, Essex (Chelmsford 400200)

Various electrical components

Bi-Pre-Pak Ltd, 222 West Road, Westcliff-on-Sea, Essex SS0 9DF (Southend 46344) Electronic parts

J. Bull (Electrical) Ltd, 7 Park Street, Croydon CR0 1YD Relays, electronic parts, transformers

Field Electric Ltd, 3 Shenley Road, Borehamwood, Herts Relays, electronic parts

Home Radio, 234 London Road, Mitcham, London Electronic parts

C. W. Wheelhouse Ltd, 11 Bell Road, Hounslow, Middlesex (01–570–3501) Relays, motors, various electrical supplies

Engineering supplies

K. R. Whiston, New Mills, Stockport SK12 4PT (0663 42028) Wide range of engineers' stock, bearings etc

Glass fibre and resins

W. David & Sons Ltd, Derbyshire House, St Chad's Street, London WC1H 8AH Isopon resins and glass fibre

Glasplies, 2 Crowland Street, Southport, Lancs (Southport 40626)

Strand Glass Co Ltd, Brentway Trading Estate, Brentford, Middlesex (01–568–9517)

Index

Make Electricity—Index 1
Acid, sulphuric, 95, 116
Agriculture, power requirements, 23
Air conditioning, power requirements, 22
Alkaline batteries, 95–6
Alternating current (ac), definition, 19
Alternator, 24–5
  advantages over dynamo, 29
  servicing, 31–3
  specifications, 30–1
Altitude, effects on engine efficiency, 46–7
Ampere (unit), 11–12
Ampere-hour (unit), 89
Armature, 25
  rewinding, 39–40

Battery, alkaline, 95–6
  comparison table, 97
  dry, 87
  faults, 94
  lead/acid, 88–95
  lithium/chlorine, 98
  lithium/sulphur, 98
  maintenance and testing, 92–3
  silver/zinc, 96–7
  sodium/sulphur, 98
  storage, 86–99
  traction, 90
Beatson (Robert), windmill design, 66
Bicycle wheel, as windmill vane, 61
Blanket, electric, power requirements, 22
Boiler fan, power requirements, 22
British Thermal Unit (BTU), 16
Brooder, power requirements, 23

Cable, current rating, 121
  overhead, safe erection, 115
Cadmium sulphide photocell, 75–6
Central-heating pump, power requirements, 22
Changeover circuit, 109
  switches, 108
Commutator, 26
  renovating, 35–6
Converter, 100–7
  rotary, 101–2
Current ratings, for cables, 121
Cut-in speed, windmill, 65
Cut-out, for dc generators, 34
  adjustment, 36
  relay as, 40

Deep freeze, power
requirements, 22
Direct current (dc),
definition, 19
Dishwasher, power
requirements, 22
Dynamo (dc generator),
rewiring as alternator, 37–43
servicing, 35–7
theory, 26–7
Dynamotor, 101–2

Edison (Thomas), alkaline
battery, 95
Electric shock, effects, 111–14
Electro magnet, for
generator, 28

Fan, boiler, power
requirements, 22
car, as windmill, 62
table, power
requirements, 22
Faraday, Michael, 24–5
Faure storage battery, 89
Feed grinding, power
requirements, 23
Field coils, generator, 28,
36, 40
Field, residual, 28
restoring in old
generator, 36
Floor polisher, power
requirements, 22
Food mixer, power
requirements, 22
Francis (James), water
turbine, 71–2
Frequency (ac), 19, 42
Fuel requirements of
generators, 46–7

Generating sets, commercial,
45–8
fuel consumption, 46–7
Generator, ac, theory, 25–6
commercial types, 43–4
dc, theory, 26–7
revolving field, 27
Glass-reinforced plastic (GRP)
for construction, 63–4
Grain drying, power
requirements, 23
elevating, power
requirements, 23

Hair dryer, power
requirements, 22
Hay curing, power
requirements, 23
Hertz (unit), 19
Honda Ltd, generating
sets, 45–6
Horizontal-shaft windmills,
52–65
Horsepower, 14
relationship to wattage, 14,
48
Hydraulic ram, 73–4
Hydroelectric power, 71–4
Hydrometer, use for testing
batteries, 92–3

Imperial Wire Gauge (swg),
117–18
Incubator, power
requirements, 23
Insulators, 15
Intelsat 4, 75
Inverters, 100
oscillator type, 102–7
vibrator type, 102
Iron, electric, power
requirements, 22

Kettle, electric, power
requirements, 22
Kilohm (unit), 15
Kilovolt (unit), 11
Kilowatt (unit), 13
Kilowatt-hour (unit), 14

Lead/acid battery, 88–95
Lights, power requirements, 22
Lithium/chlorine battery, 98
Lithium/sulphur battery, 98
Lucas, Joseph Ltd, generating
equipment, 30, 44
solar cells, 76–7

Magnus effect, 67
Megawatt (unit), 13
Megohm (unit), 15
Milking machine, power
requirements, 23
Milliampere (unit), 11–12
Millivolt (unit), 11
Milliwatt (unit), 13
Motors, electric, power
requirements, 23
synchronous, 42
Multivibrator circuit, 104

National Research Council
of Canada, windmill, 68
Nickel/cadmium battery, 95
Nickel/iron battery, 95
Noise of engines, suppression,
48

Ohm (Georg), resistance
law, 15
Ohm (unit), 15
Oxidation in batteries, 94

Peltier (Jean), thermoelectric
effect, 84
Pelton (Lester), water
wheel, 71
Planté (Gaston), storage
battery, 88
Potassium hydroxide, in
batteries, 95–6, 116
Power factor, 14
of generators, 46
Power requirements of
equipment, 22–3

Radio, power requirements, 22
Razor, electric, power
requirements, 22
Record player, power
requirements, 22
Refrigerator, power
requirements, 22
Resistance, electrical, 15–16
calculations, 16–18

Safety precautions, 111–16
Sail-bars, windmill, 55
Sail-stock, windmill, 55
Savonius (Sigurd), windmill
design, 66–8
Seebeck (Thomas),
thermoelectric effect, 81–2
Sewing machine, power
requirements, 22
Shock, electric, 111–14
Short-circuit, 18
Silicon photocells, 76
Silo unloading, power
requirements, 23
Silver/zinc battery, 96–7
Slip rings, ac generator, 29
manufacture, 38
windmill mountings, 59
Sodium/sulphur battery, 98
Solar panels, erecting, 78–80
power, utilisation, 75–80
Sulphation of batteries, 94
Sulphuric acid, density, 95
precautions in handling, 116
Suppliers, 123–4

Tail, windmill, design, 60
Television sets, power
requirements, 22
Temperature effects on
batteries, 91
on hydrometer readings, 93
rating of cables, 114–15

Thermocouple, 82–3
Thermoelectric generator, 84
  series, 83
Toaster, electric, power
  requirements, 22
Transformer, 20, 40
Turgo Impulse water
  turbine, 71
Typewriter, electric, power
  requirements, 22

Vacuum cleaner, power
  requirements, 22
Vertical-shaft windmills,
  65–70
Volt (unit), 10
Voltage, effects, 11
  root-mean-square, 19, 42–3

Warped sail (windmill),
  design, 54–5
Washing machines, power
  requirements, 22
Water power, 71–4
Watt (unit), 13
Wattage requirements, 22–3
Watt-hour (unit), 14, 89
Weatherproofing equipment,
  70
Wind energy, 51–2
Windmill, mounting, 69–70
  multi-blade, 60–1
  propeller, 62–4
  slat, 53–8
  vertical-shaft, 65–8
Wind speed, estimation, 53
Wire gauge, Brown and
  Sharpe (awg), 119–20
  Imperial (swg), 117–18